Methods of Resolution for Selected Boundary Problems in Mathematical Physics

Documents on Modern Physics

Edited by

ELLIOT W. MONTROLL, *University of Rochester*
GEORGE H. VINEYARD, *Brookhaven National Laboratory*
MAURICE LÉVY, *Université de Paris*

A. ABRAGAM L'Effet Mössbauer

K. G. BUDDEN Lectures on Magnetoionic Theory

J. W. CHAMBERLAIN Motion of Charged Particles in the Earth's Magnetic Field

S. CHAPMAN Solar Plasma, Geomagnetism, and Aurora

H.-Y. CHIU Neutrino Astrophysics

A. H. COTTRELL Theory of Crystal Dislocations

J. DANON Lectures on the Mössbauer Effect

B. S. DEWITT Dynamical Theory of Groups and Fields

R. H. DICKE The Theoretical Significance of Experimental Relativity

M. GOURDIN Lagrangian Formalism and Symmetry Laws

D. HESTENES Space–Time Algebra

J. G. KIRKWOOD Dielectrics – Intermolecular Forces – Optical Rotation; Macromolecules; Proteins; Quantum Statistics and Cooperative Phenomena; Selected Topics in Statistical Mechanics; Shock and Detonation Waves; Theory of Liquids; Theory of Solutions

V. KOURGANOFF Introduction to the General Theory of Particle Transfer

R. LATTÈS Methods of Resolution for Selected Boundary Problems in Mathematical Physics

J. LEQUEUX Structure and Evolution of Galaxies

F. E. LOWE Symmetries and Elementary Particles

P. H. E. MEIJER Quantum Statistical Mechanics

M. MOSHINSKY Group Theory and the Many-body Problem

M. NIKOLIĆ Analysis of Scattering and Decay

M. NIKOLIĆ Kinematics and Multiparticle Systems

A. B. PIPPARD The Dynamics of Conduction Electrons

H. REEVES Stellar Evolution and Nucleosynthesis

L. SCHWARTZ Application of Distributions to the Theory of Elementary Particles in Quantum Mechanics

J. SCHWINGER Particles and Sources

J. SCHWINGER and D. S. SAXON Discontinuities in Waveguides

M. TINKHAM Superconductivity

Methods of Resolution for Selected Boundary Problems in Mathematical Physics

ROBERT LATTÈS

Groupe SEMA, Paris

GORDON AND BREACH Science Publishers

New York   London   Paris

Editorial office for the United Kingdom:

Gordon and Breach Science Publishers Ltd.
12 Bloomsbury Way
London W.C.1.

Editorial office for France:

Gordon & Breach
7–9 rue Emile Dubois
Paris 14ᵉ

Distributed in Canada by:

The Ryerson Press
299 Queen Street West
Toronto 2B, Ontario

Documents on Modern Physics

Seventy years ago, when the fraternity of physicists was smaller than the audience at a weekly physics colloquium at a major university, a J. Willard Gibbs could, after ten years of thought, summarize his ideas on a subject in a few monumental papers or in a classic treatise. His competition did not intimidate him into a muddle of correspondence with his favorite editor nor did it occur to his colleagues that their own progress was retarded by his leisurely publication schedule.

Today the dramatic phase of a new branch of physics spans less than a decade and subsides before the definitive treatise is published. Moreover, modern physics is an extremely interconnected discipline, and the busy practitioner of one of its branches must keep aware of breakthroughs in other areas. An expository literature which is clear and timely is needed to relieve him of the burden of wading through tentative and hastily written papers scattered in many journals.

To this end, we have undertaken the editing of a series of books, entitled *Documents on Modern Physics*, which will make available selected reviews, lecture notes, conference proceedings, and important collections of papers in branches of physics of special current interest. Complete coverage of a field will not be the primary aim. Rather, we shall emphasize readability, speed of publication, and importance to students and research workers. The books will appear in low-cost paperback editions, as well as in clothbound editions. The scope will be broad, and the style informal.

From time to time, older branches of physics come alive again, and forgotten writings acquire relevance to recent developments. We expect to make a number of such works available by including them in this series along with new works.

ELLIOT W. MONTROLL
GEORGE H. VINEYARD
MAURICE LÉVY

Preface

In this book, the author has gathered together a series of lectures given during 1960 and 1961 at the Faculty of Sciences in Paris, under the title "Applied Mathematical Physics".

It is compounded therefore, on the one hand, of various original presentations or works and, on the other hand, in an evident regard for coherence, of various elements drawn from specialized works and, above all, articles. The author must therefore thank here all those who have made involuntary contributions to this work.

Finally, the following point should be emphasized: the purpose intended was essentially to provide those listening to the lectures (students or specialized systems engineers) with working tools and to show by means of definite, actual examples how these tools are used. That is why, in view of the time alloted to the course, rigor has often had to be sacrificed and proofs of theorems omitted in the interest of a heuristic treatment.

ROBERT LATTÈS

Contents

Part one

SPECTRAL METHODS

INTRODUCTION

Linear Operators

DEFINITIONS

Let E be a given vector space; we will call any mapping of a subset C_L of E into a subset Γ_L of E an *operator* L; this correspondence is also called an *endomorphism* of E. To each $x \in C_L$ there corresponds a $y \in \Gamma_L$. We write:

$$y = Lx$$

C_L is the *domain of definition* of the operator L.

Γ_L is the *range*.

If L maps E onto E we have an *isomorphism*.

An operator L is said to be *linear* if it satisfies the following conditions:

1) C_L is a vector subspace of E

2) $\forall x_1$ and $x_2 \in C_L$ we have $L(x_1 + x_2) = Lx_1 + Lx_2$

3) $\forall x \in C_L$ we have $L(\alpha x) = \alpha Lx$

where α is a real or complex scalar.

(2) and (3) imply that Γ_L is also a vector subspace of E.

We mention, as examples, the operators defined on the space of the real-valued functions $\varphi(S)$

$$\int_b^a K(s, t)\, \varphi(t)\, dt, \quad \frac{d}{ds} \varphi(S), \quad S\varphi(S)$$

GENERAL REMARK

This course is mainly devoted to the solution of equations of type

$$L\varphi = f \tag{1}$$

where f is a given function, L an operator defined on the space of these functions, φ an unknown function. Two methods are possible.

1) If there exists an operator L^{-1} such that:

$$\varphi = L^{-1}f$$

solving (1) reduces to finding L^{-1}. If L is a differential operator, L^{-1} is an integral operator whose kernel is called Green's function. See *Methods of Mathematical Physics*, by Friedmann.

2) We look for the *eigenfunctions* of L_L such that

$$L\psi_n = \lambda_n\psi_n$$

where λ_n is a scalar (called the eigenvalue of L)

If these functions span all the space considered we can write

$$f = \sum a_n\psi_n$$

For instance, if the system ψ_n is orthonormal, we get:

$$a_n = (f, \psi_n)$$

Assuming we can expand φ in the form, $\sum b_n\psi_n$, we deduce:

$$L\varphi = L(\sum b_n\psi_n) = \sum b_n L\psi_n = \sum b_n\lambda_n\psi_n$$

whence we deduce, by identification,

$$\varphi = \sum \frac{(f, \psi_n)}{\lambda_n}\psi_n$$

This second method is based on the spectral theory of the operators L and their representations. In fact, in applications, there arise many difficulties which do not occur when L is a matrix (Ex.: Is the system of eigenfunctions complete?). To refresh our memory we will study some fundamental properties of some operators and of their representations when $E = C^n$ a space with n complex dimensions.

CHAPTER 1

Introduction to the Spectral Study of the Endomorphisms of C^n

HERMITIAN SPACE

The vector space C^n endowed with the hermitian product $\langle x, y \rangle$ is called a *hermitian space*.

Hermitian product

We define the *hermitian product* of 2 vectors x and y as the composition law, denoted $\langle x, y \rangle$, defined by the following properties:

$$\langle \alpha x, y \rangle \quad = \alpha \langle x, y \rangle$$

$$\langle x, \alpha y \rangle \quad = \bar{\alpha} \langle x, y \rangle$$

$$\langle x + y, z \rangle = \quad \langle x, z \rangle + \langle y, z \rangle \quad \text{left distribution}$$

$$\langle x, y + z \rangle = \quad \langle x, y \rangle + \langle x, z \rangle \quad \text{right distribution}$$

If (x, y) is the cartesian product in C^n we get $\langle x, y \rangle = (x, \bar{y})$, which implies that $\langle y, x \rangle = \overline{\langle x, y \rangle}$ and $\langle x, x \rangle = \overline{\langle x, x \rangle}$ is hence real.

If the space is a basis system $U = (u_1, u_2, ..., u_x)$ and if we set

$$x = x^i u_i \quad y = y^j u_j$$

(following Einstein's convention) we can write

$$\langle x, y \rangle = \langle x^i u_i, y^j u_j \rangle = x^i \overline{y^j} \langle u_i, u_j \rangle$$

Two vectors are orthogonal if $\langle x, y \rangle = 0$. Two real vectors are hence orthogonal for the scalar product and the hermitian product since C^n admits real orthonormal bases for the scalar product and orthonormal bases for the hermitian product.

If U is orthonormal,

$$\langle u_i, u_j \rangle = \delta_{i,j}$$

(where $\delta_{i,j}$ is the Kronecker symbol) the basis is said to be *orthonormal*. Then

$$\langle x, y \rangle = \sum x^j \bar{y}^j \tag{2}$$

in particular

$$\langle x, x \rangle = \sum x^i \bar{x}^i = \sum |x^i|^2$$

Fundamental property $\langle x, x \rangle$ is always ≥ 0 and $\langle x, x \rangle = 0$ implies $x = 0$
$\langle x, x \rangle$ hence defines a norm on the space. This is why we introduce the hermitian product. There are no more isotropic elements, that is such that

$$\langle x, x \rangle = 0 \quad \text{with} \quad x \neq 0$$

Remark It follows from (2) that

$$\langle x, y \rangle = (x, \bar{y})$$

Whence we deduce

$$\langle y, x \rangle = (y, \bar{x}) = (x, \bar{y}) = \overline{\langle x, y \rangle}$$

The hermitian product is *not commutative*.

Definition of endomorphism

We call an *endomorphism* of C^n any linear mapping L of C^n *into* itself and we have

$$L(x + y) = Lx + Ly$$

$$L(\alpha x) \quad = \alpha L(x)$$

Matrix induced by an operator

Let C^n be an n-dimensional hermitian space, L an endomorphism of C^n and U a chosen basis.
    Since

$$x = x^i u_i$$

$$y = Lx$$

can also be written

$$y = x^i L u_i$$

suppose that

$$Lu_i = \alpha_i^j u_j \tag{3}$$

$$y = x^i \alpha_i^j u_j = y^j u_j \quad \text{Hence} \quad y^j = \alpha_i^j x^i$$

The matrix

$$\{\alpha_j^i\} = \begin{pmatrix} \alpha_1^1 & \alpha_2^1 & \dots & \alpha_n^1 \\ \cdot & \cdot & \cdot & \cdot & \cdot \\ \alpha_1^n & \alpha_2^n & \dots & \alpha_n^n \end{pmatrix}$$

is said to be *induced by the operator L, represented in the basis U.* It will be denoted by L_u.

Since

$$y = Lx = \{\alpha_j^i\} \begin{pmatrix} x_1 \\ \vdots \\ x_n \end{pmatrix}$$

we can consider y as the product of the induced matrix and the column vector x.

Fundamental remark.   If we consider the relation (3) as defining a change of basis substituting the vectors $u_i = Lu_i$ for the vectors, u_i the transformation matrix is:

$$\{\alpha_i^j\} = \begin{pmatrix} \alpha_1^1 & \alpha_1^2 & \dots & \alpha_1^n \\ \cdot & \cdot & \cdot & \cdot & \cdot \\ \alpha_n^1 & \alpha_n^2 & \dots & \alpha_n^n \end{pmatrix}$$

It is the transpose of the matrix induced by the operator L. This expresses the well-known property of contravariance of the vectors x. In L_u, the ith column represents the components of the vector Lu_i. We will often make use of this remark.

Regular endomorphism

Let U and V be two distinct basis systems in C^n. Any endomorphism which transforms C^n into itself is said to be *regular.* The following property can be proved without difficulty:

An endomorphism is regular if and only if it transforms a basis into another basis.

Let L be a regular endomorphism transforming the basis U into the basis LU. The endomorphism, which is also regular, which allows us to go back from LU to U is called the *inverse* of L. We denote it by

$$(L_u)^{-1}$$

If

$$L_u = \{\alpha_j^i\}$$

we know that

$$(L_u)^{-1} = \{\beta_i^j\} = \left\{\frac{A_i^j}{\Delta}\right\}$$

where A_i^j is the minor endowed with the suitable sign of the element α_i^j of the transposed matrix of L_u and where Δ is the determinant of this matrix.

We obviously have

$$L_u L_u^{-1} = I$$

Algebra of the endomorphisms

If L and M are two endomorphisms, we can define their sum and their product:

$$S(x) = L(x) + M(x)$$

which we will denote simply by

$$S = L + M$$

$$P(x) = M[L(x)]$$

$$P = ML$$

Similary if $R(x) = \alpha L(x)$ we will write

$$R = \alpha L$$

In a basis U, $L + M$, ML and αL induce sum and product matrices.

Then let L be an operator defined on C^n, and let L_u and L_v be its induced matrices in the basis systems U and V. There exists a regular endomorphism which associates the basis V to the basis U—more precisely, the vector v_i to the vector u_i— . We denote it by C_{vu} and let $(C_{vu})_u$ be its induced matrix in the basis U. The operator C_{vu}^{-1} transforms the basis V into the basis U; denote it by C_{uv}.

Then let x and y be vectors in C^n transformed by the operator L. Referred to the basis U, by means of the operator C_{uv}, they become respectively $C_{uv}x$ and $C_{uv}y$, and the coordinates of these last vectors with respect to the basis U are those of x and y with respect to the basis V. Thus in the space C^n

referred to the basis U

$$(C_{uv})_u y = L_v (C_{uv})_u x$$

but in the system with basis U

$$y = L_u x$$

Hence

$$L_v (C_{uv})_u x = (C_{uv})_u L_u x$$

Multiplying to the right the two sides of the above relation by $(C_{vu})_u$ we obtain the fundamental relation

$$\boxed{L_v = (C_{uv})_u L_u (C_{vu})_u} \qquad (4)$$

SPECIAL ENDOMORPHISMS DEFINED IN C^n

Adjoint of an operator L

This can be defined because of the hermitian product. It is the operator L' defined by the relation

$$\langle L'x, y \rangle = \langle x, Ly \rangle \quad \text{for all} \quad x, y \in C^n$$

L' induces in U a matrix L'_u which is the adjoint of L_u and we have (α'_{ji}) = $(\bar{\alpha}_{ij})$

Indeed: Let $\{\alpha'^i_j\}$ be the matrix induced by the operator L'

$$\langle L'x, y \rangle = \langle x^i L'u_i, y^j u_j \rangle$$
$$= \langle x^i \alpha'^j_i u_j, y^j u_j \rangle$$
$$= x^i \bar{y}^j \alpha'^j_i$$

Similarly

$$\langle x, Ly \rangle = \langle x^i u_i, y^j Lu_j \rangle$$
$$= \langle x^i u_i, y^j \alpha^i_j u_i \rangle$$
$$= x^i \bar{y}^j \bar{\alpha}^i_j$$

whence, by identification,

$$\alpha'^j_i = \bar{\alpha}^i_j \qquad (5)$$

The following relations are proved without difficulty

$$(L + M)' = L' + M'$$
$$(\alpha L)' \quad = \bar{\alpha} L'$$
$$(ML)' \quad = L'M'$$

2*

with, of course,

$$(L')' = L \quad \text{and} \quad (L^{-1})' = (L')^{-1}$$

Hermitian operator

By definition, an operator is said to be *hermitian* if it coincides with its adjoint

$$L = L'$$

It follows, from (5), that the α_i^j are real and $\alpha_i^j = \bar{\alpha}_j^i$.

Examples

1) The operator L such that

$$LQ = \int_b^a K(s, t)\, Q(t)\, \mathrm{d}t$$

is hermitian if the kernel $K(s, t)$ satisfies the condition

$$K(s, t) = \overline{K(t, s)}$$

Indeed:

$$\langle LQ, \psi \rangle = \int_a^b \int_a^b K(s, t)\, Q(t)\, \overline{\psi(s)}\, \mathrm{d}t\, \mathrm{d}s$$

$$= \int_a^b \int_a^b K(t, s)\, Q(s)\, \overline{\psi(t)}\, \mathrm{d}s\, \mathrm{d}t$$

$$= \int_a^b \int_a^b \overline{K(s, t)}\, \overline{\psi(t)}\, Q(s)\, \mathrm{d}s\, \mathrm{d}t = \langle Q, L\psi \rangle$$

2) The operator L such that $LQ = \dfrac{1}{2\pi i}\dfrac{\mathrm{d}}{\mathrm{d}s} Q(s)$ is hermitian if the function Q vanishes at the boundaries a and b of the interval of definition. Indeed

$$\langle LQ, \psi \rangle = \int_a^b \frac{1}{2\pi i}\left[\frac{\mathrm{d}}{\mathrm{d}s} Q(s)\right] \cdot \overline{\psi(s)}\, \mathrm{d}s$$

Integrating by parts:

$$\langle LQ, \psi \rangle = -\frac{1}{2\pi i} \int_a^b Q(s) \frac{\mathrm{d}}{\mathrm{d}s} \overline{\psi(s)}\, \mathrm{d}s = \langle Q, L\psi \rangle$$

3) *Skew-hermitian operators*   An operator is said to be skew-hermitian if its adjoint is equal to its negative.

$$L' = -L$$

From (5) we now deduce

$$\alpha^j_i = -\bar{\alpha}^i_j$$

The coefficients α^i_i are pure complex.

Remark.   If L is hermitian, iL is skew-hermitian and vice versa.

4) *Theorem*   Every endomorphism is the sum of a hermitian endomorphism and a skew-hermitian endomorphism.

Indeed, decompose L by separating real and imaginary parts

$$L = B + iC$$

with

$$\alpha^j_i = \beta^j_i + i\gamma^j_i$$

We deduce

$$\alpha'^j_i = \bar{\alpha}^i_j = \beta^i_j - i\gamma^i_j$$

We then form

$$\theta^j_i = \left(\frac{L + L'}{2}\right)^j_i = \frac{\beta^j_i + \beta^i_j + i(\gamma^j_i - \gamma^i_j)}{2}$$

which implies that

$$\theta^j_i = \frac{\beta^i_j + \beta^j_i + i(\gamma^i_j - \gamma^j_i)}{2} = \theta^i_j$$

$$\theta = \frac{L + L'}{2} \quad \text{is hermitian.}$$

Similarly we form

$$\eta^j_i = \left(\frac{L - L'}{2}\right)^j_i = \frac{\beta^j_i - \beta^i_j + i(\gamma^j_i + \gamma^i_j)}{2}$$

whence

$$\bar{\eta}^j_i = -\eta^i_j \qquad \eta = \frac{L - L'}{2}$$

is skew-hermitian
and since:

$$L = \frac{L + L'}{2} + \frac{L - L'}{2}$$

the theorem is proved.

5) *Normal endomorphism*   An endomorphism L is said to be normal if it commutes with its adjoint, that is if

$$LL' = L'L$$

If we set

$$L = B + iC$$

$$L' = B - iC$$

we deduce that

$$LL' = B^2 + C^2 + i(CB - BC)$$

$$L'L = B^2 + C^2 - i(CB - BC)$$

whence the necessary and sufficient condition

$$\boxed{CB = BC}$$

EIGENSTRUCTURE OF HERMITIAN SPACES

Definitions

We wish to find all the pairs (λ, x) composed of a scalar λ and a vector x such that:

$$Lx = \lambda x \quad x \neq 0 \tag{6}$$

If x is a solution, so is the vector Kx. To remove the resulting indetermination we generally impose an additional condition called the normalization condition because we often require that $\langle x, x \rangle = 1$, which implies that

$$K = \frac{1}{\|x\|}$$

The λ and the x which satisfy (6) are respectively the *eigenvalues* and *eigenvectors* of the operator L.

Fundamental theorems

THEOREM I *Every endomorphism admits n eigenvalues, distinct or not, which are the roots of the equation*

$$|L_u - \lambda I| = 0$$

In particular

$$\det (L_u - \lambda I) = \det (L_v - \lambda I)$$

Indeed, (6) implies that, in the space C_n referred to the basis U,

$$(L_u - \lambda I) x = 0$$

a homogeneous linear system which admits a non-trivial solution only if

$$|L_u - \lambda I| = 0$$

The two following propositions are related to this theorem:

PROPOSITION 1   To each eigenvalue there corresponds *at least* one eigenvector and an eigenvector is associated with *only one* eigenvalue.

The first part follows from the definition of an eigenvector; the second
can be proved by contradiction. The relations

$$L_u x = \lambda_1 x \quad L_u x = \lambda_2 x$$

imply

$$(\lambda_1 - \lambda_2) x = 0$$

and since $x \neq 0$,

$$\lambda_1 = \lambda_2$$

PROPOSITION 2   A linear combination of eigenvectors is itself an eigenvector only if the eigenvectors of this combination belong to the same
eigenvalue, and we then have an eigenvector subspace. The corresponding
eigenvalue is a root of det $(L_u - \lambda I) = 0$ of order greater or equal to the
dimension of the generated subspace.

Proof

1) Let there be p eigenvectors C_i; λ_i are the associated eigenvalues
$(L x_i = \lambda_i x_i)$; consider the combination

$$x_0 = \sum a_i x_i$$

If x_0 is itself an eigenvector

$$L x_0 = \lambda_0 x_0$$

Now

$$L x_0 = \sum a_i L x_i = \sum a_i \lambda_i x_i$$

so that we must have

$$\sum a_i \lambda_i x_i - \lambda_0 \sum a_i x_i = 0$$

that is

$$\sum a_i (\lambda_i - \lambda_0) x_i = 0 \quad \text{for all} \quad a_i.$$

Such a relation is possible only if $a_i(\lambda_i - \lambda_0) = 0$. Whence the type of the
possible solution if we wish that the linear combination of p eigenvectors

be an eigenvector

$$a_{p+1} \cdots a_r = 0$$

$$\lambda_0 = \lambda_1 = \lambda_2 = \cdots = \lambda_p \quad \text{where} \quad a_1 \ldots a_r \neq 0$$

and this shows that the common value of these λ is a root of order at least equal to (p).

2) All the combinations of the preceding (p) vectors are eigenvectors. Hence if we add 0, we get an eigenvector subspace of dimension p.

All the minors of order $\geqq n - p$ of the determinant $(L_u - \lambda_0 I)$ are hence zero.

We then consider the determinant

$$|L_u - \lambda_0 I - \lambda I|$$

It can be expanded into

$$(-1)^n \lambda^n + (-1)^{n-1} \quad \text{Im} \, (L_u - \lambda_0 I) \, \lambda^{n-1} + \cdots + (-1)^p \sum \varDelta_{n-p} \lambda^p + \cdots$$
$$- \sum \varDelta_{n-1} \lambda + \varDelta_n$$

denoting by $\varDelta_{n-p}$ a principal minor of the determinant $|L_u - \lambda_0 I|$.

From what came previously we see that at least the last p terms of this expansion are zero and hence the eigenequation

$$|L_u - \lambda_0 I - \lambda I| = 0$$

admits $\lambda = 0$ as a root of order at least equal to p.

THEOREM II (and definition) *If an endomorphism admits n distinct eigenvalues it admits a unique spectral basis composed of the n distinct eigenvectors associated with the eigenvalues.*

Indeed, to each eigenvalue there corresponds a unique eigenvector and these vectors are linearly independent by Proposition (2).

THEOREM III *If an endomorphism admits a spectral basis: either the eigenvalues are distinct and this basis is unique or the eigenvalues are not distinct and to each eigenvector of order p there corresponds an eigenvector subspace of dimension p, the decomposition into subspaces being unique by virtue of Proposition (2).*

THEOREM IV *To an eigenvalue of order p there correspond at most p linearly independent eigenvectors, but this number is bounded below only by 1.*

Proposition (2) shows indeed that their number is at most equal to p; but it can be less than p, as is shown by Jordan's example

$$J_p(\lambda) = \begin{vmatrix} \lambda & 1 & & & & 0 \\ & \lambda & 1 & & & \\ & & \lambda & 1 & & \\ & & & \ddots & 1 & \\ & & & & \lambda & \end{vmatrix} = \lambda^p$$

To the values $\lambda = 0$ of order p there corresponds only the eigenvector $(1, 0, \ldots, 0)$.

THEOREM III b   *If an endomorphism admits the spectral basis X, its induced matrix in this basis, D_x, is diagonal.*

$$D_x = \begin{pmatrix} \lambda_1 & & & \\ & \lambda_2 & & 0 \\ & & \ddots & \\ 0 & & & \lambda_n \end{pmatrix}$$

This follows immediately from the fact that the ith column of D_x must represent the vector $L(x_i) = \lambda_i x_i$. If L_u is the matrix induced by L in the basis U, the formula (4) with $V = X$ gives:

$$D_x = (C_{ux})_u L_u (C_{xu})_u$$

or

$$L_u = (C_{xu})_u D_x (C_{ux})_u$$

THEOREM V (Jordan)   *If an endomorphism L does not have a spectral basis, it has eigenvalues λ_i of order p_i. We can decompose C^n into a direct sum of subspaces such that in each one of them L reduces to a Jordan endomorphism, that is to say that in each one of them we can find a basis $U\alpha$ such that $\{L_u\alpha\}$ $= J_{p_\alpha}(\lambda)$.*

This fundamental theorem is proved in three steps.

1) *Reduction to upper triangular form*

$$L_{x+y} = \left(\begin{array}{c|c} D_{\lambda_i} & B_1 \\ \hline 0 & C_1 \end{array} \right) \qquad \left(\begin{array}{c|c} D_{1,\lambda_i} & B_2 \\ \hline 0 & C_2 \end{array} \right)$$

$$\text{I} \qquad\qquad\qquad \text{II}$$

Let U be a basis system for C^n; consider the subspace X generated by all the *distinct* eigen vectors of L and complete their set by adjoining certain vectors of U forming the set Y_1 so that the union $X + Y_1$ generates all the space. The matrix L_{X+Y_1} will necessarily be of the form (I) where $D\lambda_i$ is the diagonal matrix of the distinct eigenvalues completed by 0 below and to the left. The matrix C_1 defines an operator, in the subspace generated by Y_1, whose eigenvalues are eigenvalues of L with lower order. Hence we can, as previously, use X_1, the basis composed of the eigenvectors of C_1 and complete by Y_2 (vectors of Y_1); C_1 will then be of form (II) and so forth. Since these operations cannot be repeated indefinitely we will finally obtain a basis $X + X_1 + \cdots + X_k$ in which the matrix induced by L will be of form:

$$L_{X+X_1+\cdots+X_K} = \begin{pmatrix} D_{\lambda_i} & x\ x \cdots\cdots\cdots & x \\ & D_{1.\lambda_i} & \begin{matrix} x \\ \vdots \end{matrix} \\ & & \ddots & x \\ 0 & & & D_{K.\lambda_i} \end{pmatrix}$$

2) *Elimination of the components on the vectors corresponding to distinct eigenvalues* We then show that in the previous matrix we can annihilate any coefficient located at the intersection of a row and a column corresponding to distinct eigenvalues, for instance a_i^j. We have seen that the ith column of the matrix L represents the components of the vector $L(x_i)$, where x_i denotes the ith vector of the basis $X + X_1 + \cdots + X_k$. Hence

$$L(x_i)_i = \lambda_p$$
$$L(x_j)_i = a_i^j$$

$$\begin{array}{cc} L(x_i) & L(x_j) \\ \downarrow & \downarrow \\ \lambda_p \cdots\cdots\cdots\cdots a_i^j\ i\text{th row} \\ \\ \\ \uparrow & \lambda_q \\ i\text{th column} & j\text{th column} \end{array}$$

We try, by replacing X_j by a combination u_j of x_j and x_i, to annihilate the component a_{ij} of the new basis

$$u_j = x_j + \alpha x_i \quad \text{hence} \quad (u_j)_i = \alpha$$

where α is a given number, temporarily indeterminate. Since the operator L is linear

$$L(u_j) = L(x_j) + \alpha L(x_i)$$

so that

$$L(u_j)_i \equiv a_i^j + \alpha\lambda_p$$

in the basis $X + X_1 + \cdots + X_k$

With respect to the basis $x_1 \cdots x_{j-1}, u_j, x_{j+1} \cdots$, identical to the previous one except for the vector x_j, we have this time:

$$L(u_j) = (a_i^j + \alpha\lambda_p)\, x_i + \lambda_q x_j$$

that is to say
$$= (a_j^i + \alpha\lambda_p)\, x_i + \lambda_q(u_j - \alpha x_i)$$
$$= [a_i^j + \alpha(\lambda_p - \lambda_q)]\, x_i + \lambda_q u_j$$

and if we want the coefficient of x_i to be zero in this basis, we will choose α such that:

$$a_i^j + \alpha(\lambda_p - \lambda_q) = 0 \tag{7}$$

which requires that $\lambda_p - \lambda_q$ is not equal to 0. Note (cf. the example discussed in this paper) that we must annihilate the a_{ij} for increasing j and decreasing i.

Remark   As an exercise, we see that formula (7) can also be obtained by using the formula for change of basis (4) in which:

$$(C_{vu})_u = \begin{pmatrix} 1 & & & & & & & & \\ & 1 & & & & & & & \\ & & 1 & & & & & & \\ & & & 1 & \cdots\cdots\cdots\cdots & & & & \cdots \text{ith row} \\ & & & & 1 & & & & \\ & & & & & 1 & & & \\ & & & & & & 1 & & \\ & & & \alpha & \cdots\cdots\cdots & & & \ddots & \cdots \text{jth row} \quad \text{etc.} \\ & & & & & & & & 1 \end{pmatrix}$$

Hence we obtain a reduced form which contains non-zero elements only at the intersection of a row and a column associated with the same eigenvalue. We can finally decompose the space as a direct sum of subspaces in each of which we have a matrix of the form

$$L_u = \begin{pmatrix} \lambda & a_{12} & a_{13} & a_{14} \\ & \lambda & & \\ & & \lambda & \\ & & & \lambda \end{pmatrix}$$

3) We finally show that for a well-determined eigenvalue λ, L can be reduced by a change of basis to the form

$$[L_v] = \left(\begin{array}{c|c} D_\lambda & 0 \\ \hline 0 & J_\lambda \end{array} \right)$$

where D_λ is a diagonal matrix containing only distinct and simple eigenvalues λ_i and where J_λ is a Jordan matrix, that is to say of the type mentioned in Theorem IV. Indeed, suppose to fix our ideas that there is only one eigenvalue of order 4, all the other λ_i being simple. We can already reorder the vectors in such a way that L is of form

$$\left(\begin{array}{c|c} D_{\lambda_i} & B \\ \hline & \begin{array}{cccc} \lambda & c_1^2 & c_1^3 & c_1^4 \\ 0 & \lambda & c_2^3 & c_2^4 \\ & & \lambda & c_3^4 \\ & 0 & & \lambda \end{array} \end{array} \right)$$

and it follows from 2) that we can annihilate all the elements of B; we will assume this has been done. We then consider the matrix M in the right lower corner, where we can always suppose that the coefficients bordering the principal diagonal c_1^2, c_2^3, c_3^4 are $\neq 0$. Denoting by $x_1, \ldots, x_4$ the basis vectors, we get

$$L(x_4) = c_1^4 x_1 + c_2^4 x_2 + c_3^4 x_3 + x_4$$

We then set

$$u_3 = c_1^4 x_1 + c_2^4 x_2 + c_3^4 x_3$$

whence

$$L(x_4) = u_3 + x_4.$$

In the basis system x_1, x_2, u_3, x_4 the corresponding matrix M will be of form

$$\begin{pmatrix} \lambda & C_1^2 & C_1^3 & 0 \\ & \lambda & C_2^3 & 0 \\ & 0 & \lambda & 1 \\ & & & \lambda \end{pmatrix}$$

Indeed:

$$L(u_3) = C_1^4 \lambda x_1 + C_2^4(C_1^2 x_1 + \lambda x_2) + C_3^4(C_1^3 x_1 + C_2^3 x_2 + \lambda x_3)$$

$$= \lambda(C_1^4 x_1 + C_2^4 x_2 + C_3^4 x_3) + (C_2^4 C_1^2 + C_3^4 C_1^3) x_1 + C_3^4 C_2^3 x_2$$

$$= C_1'^3 x_1 + C_2'^3 x_2 + \lambda u_3, \quad C_1'^3 = C_1^2 C_1^4 + C_1^3 C_1^4$$

$$\underline{C_2'^3 = C_2^3 C_3^4 \quad \text{hence} \neq 0}$$

which allows us to calculate $C_1'^3$ and $C_2'^3$ by identification.

We then set

$$u_2 = C_1'^3 x_1 + C_2'^3 x_2 \quad \text{and} \quad LU_3 = U_2 + \lambda U_3$$

$$Lu_2 = C_1'^2 x_1 + \lambda U_2 = C_1'^3 \lambda x_1 + C_2'^3(C_1^2 x_1 + \lambda x_2)$$

$$= \lambda U_2 + C_{23}' C_{12} x_1.$$

M takes the new form

$$\begin{pmatrix} \lambda & C_1'^2 & 0 & 0 \\ & \lambda & 1 & 0 \\ & & 0 & 1 \\ & & & 0 \end{pmatrix} \quad \text{where} \quad C_1'^2 = C_1^2 C_2'^3 \neq 0$$

It suffices hence to set

$$u_1 = C_1'^2 x_1$$

$$LU_1 = \lambda U_1$$

to put M definitely in the required form.

THEOREM V b   *There exists a basis X such that*

$$L_u = (C_{xu})_u \, J_x (C_{ux})_u$$

where J_x is the Jordan form of L.

Fundamental remark   Jordan's theorem shows that there exists a unique basis for which the matrix induced by L is of type J_x. It is the *canonical form*, that is to say associated uniquely to a given endomorphism. Indeed, if L admits a multiple λ of order p and $q \leq p$ associated eigenvectors, in any Jordan form of L, the contribution of this eigenvalue will be

$$\begin{bmatrix} D_{q-1} & (\lambda) & 0 \\ & & \\ 0 & & J_{p-(q-1)} \, (\lambda) \end{bmatrix}$$

The following theorems are concerned with the adjoint endomorphism L' of a given endomorphism L.

We will need the following lemma.

The adjoint endomorphism of an endomorphism C_{vu} is $C'_{vu} = C_{v'u}$

PROPOSITION 3　Let $(C_{vu})_u$ be the matrix induced by an operator L which associates the following vectors to the basis vectors u_i

$$v_j = \alpha^i_j u_i \quad |\alpha^j_i| \neq 0$$

We have seen that we can write

$$(C_{vu})_u = \begin{pmatrix} v_1 \cdots v_n \\ \Big| \qquad \Big| \end{pmatrix} = \{\alpha^i_j\}$$

whose determinant is different from zero and

$$(C_{vu})'_u = \{\bar\alpha^j_i\} = \begin{pmatrix} \overline{v}_1 \,\rule{1.5cm}{0.4pt} \\ \vdots \quad \cdots \\ \overline{v}_n \,\rule{1.5cm}{0.4pt} \end{pmatrix}$$

whose determinant is different from zero
with

$$\overline{v}_j = \bar\alpha^i_j u_i$$

The determinant of the $\bar\alpha^j_i$ is not zero. The operator whose induced matrix is $(C_{vu})'_u$ associates to the vectors u_i the vectors v'_i, column vectors of the new matrix. Denoting by v' the set of these vectors which form a basis, we get the relation

$$(C_{vu})'_u = (C_{v'u})_u \tag{8}$$

From this formula we deduce

$$(C_{uv})'_u = (C^{-1}_{vu})'_u = (C'_{vu})^{-1}_u = (C_{v'u})^{-1}_u = (C_{uv'})_u$$

that is

$$\underline{(C_{uv})'_u = (C_{uv'})_u} \tag{8'}$$

THEOREM VI　*Let J be the Jordan form of an operator L*

a) its adjoint L' induces the Jordan form J', adjoint of J;

b) the eigenvalues of L' are conjugates of the eigen values of L;

c) if L admits a spectral basis, so does L'.

Indeed, if U is an arbitrary basis:

$$L_u = (C_{xu})_u J_x (C_{ux})_u$$

whence

$$L'_u = (C_{ux})'_u J'_x (C_{xu})'_u$$

and from (8) and (8′)

$$L'_u = (C_{ux'})_u J'_x (C_{x'u})_u$$

then let Y be a basis such that there exists an endomorphism which makes X' correspond to U in the same way as U corresponds to Y

$$(C_{ux'})_u = (C_{yu})_u \quad \text{and} \quad (C_{x'u})_u = (C_{uy})_u$$

we get:

$$\underline{L'_u = (C_{yu})_u J'_y (C_{uy})_u}$$

Which shows that in the basis Y the endomorphism L' induces J', adjoint of J.

From this form of J' we can see that if λ is an eigenvalue of L of order p with $q \leq p$ associated eigenvectors, $\bar{\lambda}$ is an eigenvalue of L' of order p with $q \leq p$ associated eigenvectors. Hence L' admits $\bar{J}$ as its Jordan form. And if L has a spectral basis, XJ is diagonal, hence so is J' and $J' = J$, which proves that Y is a spectral basis of L'.

Geometrical position of Y   The basis Y is defined by

$$C_{ux'} = C_{yu} = (C_{x'u})^{-1}$$

If

$$X = \begin{pmatrix} x_1 & . & . & x_n \\ . & . & . & . \\ . & . & . & . \\ . & . & . & . \end{pmatrix} \quad \text{and} \quad Y = \begin{pmatrix} y_1 & . & . & y_n \\ . & . & . & . \\ . & . & . & . \\ . & . & . & . \end{pmatrix}$$

From

$$C_{yu} = (C_{x'u})^{-1}$$

we get:

$$Y = (X')^{-1}$$

where X' is the adjoint matrix of X. This expresses geometrically that the two bases are complementary.

DEFINITION   *Two bases X and Y are said to be complementary if*

$$\langle x_i \bar{y}_j \rangle = \delta_{i,j},$$

the Kronecker symbol, that is to say that the matrix of the $\langle x_i \bar{y}_j \rangle$ is diagonal

$$W = \begin{pmatrix} \langle x_1, \bar{y}_1 \rangle & \langle x_1, \bar{y}_2 \rangle & \cdots & \langle x_1, \bar{y}_n \rangle \\ \langle x_2, \bar{y}_1 \rangle & \langle x_2, \bar{y}_2 \rangle & \cdots & \langle x_2, \bar{y}_n \rangle \\ \cdot & \cdot & \cdots & \cdot \\ \langle x_n, \bar{y}_1 \rangle & \langle x_n, \bar{y}_2 \rangle & \cdots & \langle x_n, \bar{y}_n \rangle \end{pmatrix} = \begin{pmatrix} x_1 \underline{\quad\quad} \\ \cdot \\ \cdot \\ x_n \underline{\quad\quad} \end{pmatrix} \begin{pmatrix} \bar{y}_1 & & \bar{y}_n \\ | & & | \\ | & & | \end{pmatrix}$$

$$= X'\bar{Y} = \bar{X}'\bar{X}'^{-1} = I$$

Q.E.D.

THEOREM VII   *If L admits a spectral basis X, the complement $(X')^{-1}$ of X is the spectral basis of L' adjoint of L. Two eigenvectors of L and L' which do not belong to conjugate eigenvalues are orthogonal.*
Let

$$LX = \lambda X \quad \text{and} \quad L'Y = \mu Y \quad \text{where} \quad \bar{\mu} \neq \lambda$$

$$\langle LX, Y \rangle = \lambda \langle X, Y \rangle$$

$$= \langle X, L'Y \rangle = \bar{\mu} \langle X, Y \rangle$$

let $(\lambda - \bar{\mu}) \langle X, Y \rangle = 0,$

since $\lambda \neq \bar{\mu}$ we must have $\langle X, Y \rangle = 0$. But this implies Theorem 8 only if all the eigenvalues are distinct, i.e. if L admits a spectral basis.

Hermitian endomorphisms   They satisfy $\underline{L - L'}$.

From Theorem (6) the eigenvalues are real.

It follows from the preceding that if X_1 and X_2 are *two eigenvectors associated with distinct eigenvalues*, they *are orthogonal.*

More precisely:

THEOREM VIII   *A hermitian endomorphism has an orthogonal spectral basis and its eigenvalues are real; conversely if an endomorphism admits an ortho-gonal spectral basis and real eigenvalues it is hermitian.*

Let λ be an eigenvalue; there exists at least one associated eigenvector X_1. For any vector Y orthogonal to X_1:

$$\langle X_1, Y \rangle = 0$$

we have $\langle X_1, LY \rangle = \langle LX_1, Y \rangle \quad \text{(since } L' = L)$

$$= \lambda_1 \langle X_1, Y \rangle = 0$$

hence LY is orthogonal to X_1

For these vectors Y, L is hence the sum of an endomorphism L_{X_1} (homothety on X_1 with magnification λ_1) and of an endomorphism L_{V_1} operating on a subspace V_1 orthogonal to X_1.

L_{V_1} is hermitian; indeed, if Y and Z are two vectors of V_1

$$\langle Y, L_{V_1}Z \rangle = \langle Y, LZ \rangle \quad \text{since} \quad \langle Y, L_{X_1}Z \rangle = 0 \quad \text{(orthogonality)}$$

and because of hermicity

$$= \langle LY, Z \rangle = \langle L_{X_1}Y, V \rangle \times \langle L_{V_1}Y, Z \rangle$$
$$= \langle L_{V_1}Y, Z \rangle$$
$$\text{since} \quad \langle L_{X_1}Y, Z \rangle = k\langle X_1, Z \rangle = 0$$

Hence $\qquad \langle Y, L_{V_1}Z \rangle \quad = \langle L_{V_1}Y, Z \rangle$

and L_{V_1} is hermitian.

The reasoning is repeated for L_{V_1} as for L, thus using up all the eigenvalues and the eigenvectors, since each time we decrease by one the dimension of the space we consider. Whence the first statement.

From Theorem VII, if L admits a spectral basis, the complement of this basis is a spectral basis of L'.

Hence L' has the same spectral basis as L and the same expression in this basis; $L' = L$; whence the converse statement.

COLLORARY VIII'   A skew-hermitian endomorphism admits an orthogonal spectral basis and pure complex eigenvalues; and conversely, if an operator A admits an orthogonal spectral basis and pure complex eigenvalues it is skew-hermitian.

This follows from Theorem VIII, since the operator iA admits a spectral basis and real eigenvalues; iA is hermitian, hence A is skew-hermitian.

STUDY OF AN EXAMPLE

Consider the integral operator

$$g(t) = \int_0^1 r(t - \tau) f(\tau) \, d\tau,$$

a convolution product denoted by $g(t) = r * f$

where $\qquad r(h) = \int_{-\infty}^{+\infty} e^{ih\omega} \varrho(\omega) \, d\omega \quad \text{with} \quad \varrho(\omega) = \frac{1}{1 + \omega^2}$

Calculation of $r(h)$

$$r(h) = \int_{-\infty}^{+\infty} e^{ih\omega} \frac{1}{1 + \omega^2} \, d\omega$$

if $\omega = \alpha + i\beta$, we have:

$$r(h) = \int_{-\infty}^{+\infty} e^{-h\beta} e^{ih\alpha} \frac{1}{1 + \omega^2} \, d\omega$$

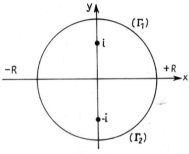

Figure 1

Depending on the sign of h we must take β positive or negative.

FIRST CASE $h > 0$ On the contour composed of the real axis $(-R, +R)$ and the semi-circle (Γ_1) with positive ordinate and radius R

$$\int_{-R}^{+R} e^{ih\omega} \frac{1}{1 + \omega^2} \, d\omega + \int_{\Gamma_1} e^{-h\beta} e^{ih\alpha} \frac{1}{1 + \omega^2} \, d\omega = 2i\pi A_i$$

where A_i is the residue of the function $\dfrac{e^{ih\omega}}{1 + \omega^2}$ at the pole i. The first integral is that which we calculate for $R \to \infty$. The second tends to 0 when $R \to \infty$. Hence:

$$\int_{-\infty}^{+\infty} e^{ih\omega} \frac{1}{1 + \omega^2} \, d\omega = 2i\pi A. \text{ The function } \frac{e^{ih\omega}}{1 + \omega^2} = \frac{e^{ih\omega}}{\omega + i} \times \frac{1}{\omega - i} \text{ has}$$

residue

$$A_i = \frac{e^{-h}}{2i} \quad \text{and} \quad \int_{-\infty}^{+\infty} \frac{e^{ih\omega}}{1 + \omega^2} \, d\omega = \pi e^{-h}$$

SECOND CASE $h < 0$ We integrate as previously, but on the semi-circle Γ_2 with negative ordinate $\omega = \alpha + i\beta$ $(\beta < 0)$

we get $$\int_{-R}^{+R} e^{ih\omega} \frac{1}{1 + \omega^2} \, d\omega + \int_{\Gamma_2} e^{-h\beta} e^{ih\alpha} \frac{1}{1 + \omega^2} \, d\omega = -2i\pi A_{-i}$$

where A_{-i} is the residue of the function $\dfrac{e^{ih\omega}}{1 + \omega^2}$ at the pole $-i$. When $R \to \infty$ the first integral tends to that which we are calculating, the second tends to 0, and we get:

$$\int_{-\infty}^{+\infty} e^{ih\omega} \frac{1}{1 + \omega^2} = -2i\pi A_{-i}$$

$$A_{-i} = \left(\frac{e^{ih\omega}}{\omega - i} \right)_{\omega = -i} = -e^{+h} \quad \text{and} \quad \int_{-\infty}^{+\infty} \frac{e^{ih}}{1 + \omega^2} = \pi e^{+h}$$

For arbitrary h we can combine these two results:

$$\int_{-\infty}^{+\infty} e^{ih\omega} \frac{1}{1 + \omega^2} \, d\omega = \pi e^{-|h|}$$

The operator $D = r*$ is a positive definite hermitian operator; the eigenvalues can hence only be positive

a) *It is hermitian*

$$\langle f, D\psi \rangle = \langle Df, \psi \rangle$$

for all functions f and ψ.

Indeed,

$$\langle f, D\psi \rangle = \int_0^1 f(t) \, dt \int_0^1 r(t - \tau) \, \psi(\tau) \, d\tau$$

$$= \int_0^1 \int_0^1 r(t - \tau) \, \psi(\tau) f(t) \, d\tau \, dt = \int_0^1 \psi(\tau) \, d\tau \int_0^1 r(t - \tau) f(t) \, dt$$

now

$$r(h) = \int_{-\infty}^{+\infty} e^{-ih\omega} \frac{1}{1 + \omega^2} \, d\omega = \int_{-\infty}^{+\infty} e^{i(-h)\omega} \frac{1}{1 + \omega^2} \, d\omega = r(-h)$$

3*

hence
$$r(t - \tau) = r(\tau - t)$$
and
$$\langle f, D\psi \rangle = \int_0^1 \psi(\tau) \, d\tau \int_0^1 r(\tau - t) f(t) \, dt = \langle Df, \psi \rangle$$

b) *It is positive definite* if $\langle f, Df \rangle$ is positive for every non-zero function f, and zero only if $f = 0$.
$$\langle f, Df \rangle = \int_0^1 \int_0^1 \pi e^{-|t-\tau|} f(t) f(\tau) \, dt \, d\tau$$
is essentially positive.

On the other hand if f an is eigenfunction
$$\langle f, Df \rangle = \langle f, sf \rangle = s\langle f, f \rangle$$
is positive which shows that *the eigenvalues* are *all positive and real*.

Calculate the second derivative

$$g''(t) = \int_0^1 r''(t - \tau) f(\tau) \, d\tau = r'' * f$$

Calculation of $r''(h)$ where $r = \tau e^{-|h|}$.
We introduce the Heaviside function
$$y(x) = \begin{cases} 1 \text{ for } x > 0 \\ 0 \text{ for } x < 0 \end{cases}$$
non-defined for $x = 0$; then
$$|h| = h[y(h) - y(-h)]$$

The derivative of $y(x)$ is the Dirac function
$$\delta(x) = \begin{cases} = 0 \text{ for } x \neq 0 \\ = \delta \text{ (Dirac measure) for } x = 0 \end{cases}$$

Thus we get:
$$[|h|]' = [h(y(h) - y(-h))]' = y(h) - y(-h) + h[\delta(h) + \delta(-h)]$$

The second term is zero, for all h, since the term in brackets is zero for $h \neq 0$ there remains
$$[|h|]' = y(h) - y(-h)$$

Then

$$(|h|)'' = \delta(h) + \delta(-h) = \begin{cases} = 0 & \text{for } h \neq 0 \\ = 2 & \text{for } h = 0 \end{cases}$$

Then

$$r' = -\pi e^{-|h|}(y(h) - y(-h))$$

$$r'' = \pi e^{-|h|}[(y(h) - y(-h))^2 - (\delta(h) + \delta(-h))]$$

now

$$[y(h) - y(-h)]^2 = 1$$

for all h

$$\delta(h)\, e^{-|h|} = \delta(-h)\, e^{-|h|} = \delta$$

Hence

$$r'' = r - 2\pi\delta$$

(Readily calculated from the distribution derivatives, cf. L. Schwartz) We then get:

$$g''(t) = r'' * f = (r - 2\pi\delta) * f = r * f - 2\pi\delta * f$$

Our search for the eigenelements leads to finding the solution of the characteristic equation:

$$r * f = sf$$

Differentiating:

$$r'' * f = sf''$$

Let

$$sf'' = r * f - 2\pi\delta * f$$

Since

$$r * f = sf$$

$$sf'' = sf - 2\pi\delta * f = sf - 2\pi f = (s - 2\pi)f$$

because the convolution product $\delta * T = T$.
  We are thus lead to

$$s_n f'' = (s_n - 2\pi)f$$

a second-order equation whose solutions are of the form

$$a_n e^{\varrho_n \tau} + b_n e^{-\varrho_n \tau} \quad \text{if} \quad s_n \neq 2\pi$$

where $\quad \varrho_n^2 = \dfrac{s_n - 2\pi}{s_n} = 1 - \dfrac{2\pi}{s_n}$, $\quad$ that is $\quad s_n = \dfrac{2\pi}{1 - \varrho_n^2}$.

We discard the solutions which have no direction $\varrho_n = \pm 1$ (which would imply that s_n is infinite). ϱ_n will hence be pure complex, or real and <1 since the operator is hermitian and positive definite and the s_n are hence real and positive.

We calculate

$$r * e^{\varrho\tau} = \int_0^1 r(t-\tau) e^{\varrho\tau} \, d\tau = \pi \int_t^1 e^{-|t-\tau|} e^{\varrho t} \, d\tau$$

Since $0 < t < 1$

$$\int_0^1 e^{-|t-\tau|} e^{\varrho\tau} \, d\tau = \int_0^t e^{-t+\tau} e^{\varrho\tau} \, d\tau + \int_t^1 e^{t-\tau} e^{\varrho\tau} \, d\tau$$

$$= \left[e^{-t} \frac{e^{(\varrho+1)\tau}}{\varrho+1} \right]_0^t + \left[\frac{e^t e^{(\varrho-1)\tau}}{\varrho-1} \right]_t^1 = \frac{e^{\varrho t}}{\varrho+1} - \frac{e^{-t}}{\varrho+1} + \frac{e^t e^{\varrho-1}}{\varrho-1} - \frac{e^{\varrho t}}{\varrho-1}$$

$$= e^t \frac{e^{\varrho-1}}{\varrho-1} - \frac{e^{-t}}{\varrho+1} - \frac{2}{\varrho^2-1} e^{\varrho t}$$

and:

$$r * e^{\varrho\tau} = \pi \left(-\frac{2}{\varrho^2-1} e^{\varrho t} + \frac{e^{t-1} e^{\varrho}}{\varrho-1} \frac{e^{\varrho}}{\varrho-1} - \frac{e^{-t}}{\varrho+1} \right)$$

Then:

$$r * (a_n e^{\varrho_n \tau} + b_n e^{\varrho_n \tau}) = \frac{2\pi}{1 - \varrho_n^2} [a_n e^{\varrho_n t} + b_n e^{-\varrho_n t}]$$

$$+ \pi e^{t-1} \left(\frac{a e^{\varrho_n}}{\varrho_n - 1} - \frac{b e^{-\varrho_n}}{\varrho_n + 1} \right)$$

$$- \pi e^{-t} \left(\frac{a}{\varrho_n + 1} - \frac{b}{\varrho_n - 1} \right)$$

and:

$$sf = s(a_n e^{\varrho_n t} + b_n e^{-\varrho_n t})$$

In order that the equation be satisfied we must have:

$$r * f = s_n f$$

$$\left. \begin{array}{c} \dfrac{a_n e^{\varrho_n}}{\varrho_n - 1} - b_n \dfrac{e^{-\varrho_n}}{\varrho_n + 1} = 0 \\[3mm] \dfrac{a_n}{\varrho_n + 1} - \dfrac{b_n}{\varrho_n - 1} = 0 \end{array} \right\} \quad \dfrac{A_n}{b_n} = e^{-2\varrho_n} \dfrac{\varrho_n - 1}{\varrho_n + 1} = \dfrac{\varrho_n + 1}{\varrho_n - 1}$$

Whence we deduce the eigenvalue equation

$$e^{-\varrho_n} = \varepsilon \frac{\varrho_n + 1}{\varrho_n - 1} \qquad (\varepsilon = \pm 1)$$

This equation shows that: ϱ cannot be real, because

$$\text{if } \varrho < 0, \quad |e^{-\varrho}| > 1 \quad \text{and} \quad \left| \frac{\varrho + 1}{\varrho - 1} \right| < 1$$

$$\text{if } \varrho > 0, \quad |e^{-\varrho}| < 1 \quad \text{and} \quad \left| \frac{\varrho + 1}{\varrho - 1} \right| > 1$$

which is impossible in each case

Hence: $\qquad\qquad\qquad\qquad \varrho_n = iV_n \qquad\qquad\qquad$ where V_n is real

and:

$$S_n = \frac{2\pi}{1 - \varrho_n^2} = \frac{2\pi}{1 + V_n^2}$$

where the eigenvalue equation is $e^{-iv} = \varepsilon \dfrac{iv + 1}{iv - 1}$

Determination of the eigenfunctions

$$f_n = a_n e^{\varrho_n t} + b_n e^{-\varrho_n t} = a_n e^{iv_n t} + b_n e^{-iv_n t}$$

$$= (a_n + b_n) \cos V_n t + i(a_n - b_n) \sin V_n t$$

where:

$$\frac{a_n}{\varrho_n + 1} = \frac{b_n}{\varrho_n - 1} = \frac{a_n + b_n}{2\varrho_n} = \frac{a_n - b_n}{2}$$

Say:

$$\frac{a_n + b_n}{2iV_n} = \frac{a_n - b_n}{2} = -i\frac{\lambda}{2}$$

where λ is an arbitrary constant; whence

$$a_n + b_n = \lambda V_n$$

$$a_n - b_n = i\lambda$$

and:

$$f_n(t) = \lambda[V_n \cos V_n t + \sin V_n t]$$

with V_n a solution of the equation:

$$e^{iV_n} = \varepsilon \frac{iV_n - 1}{iV_n + 1}$$

Let:

$$\cos V_n + i \sin V_n = \varepsilon \frac{V_n^2 - 1 + 2iV_n}{1 + V_n^2}$$

or:

$$\cos V_n = \varepsilon \frac{V_n^2 - 1}{V_n^2 + 1} \qquad \sin V_n = \frac{2V_n}{V_n^2 + 1}$$

Let:

$$\tan V_n = \frac{2V_n}{V_n^2 - 1}$$

The equation which is written as:

$$V_n^2 - 2 \cot V_n \cdot V_n - 1 = 0$$

which can be written:

$$\left(V_n - \cot \frac{V_n}{2}\right)\left(V_n + \tan \frac{V_n}{2}\right) = 0$$

The solutions are those of the equations

$$V_n = \cot \frac{V_n}{2} \quad \text{and} \quad V_n = -\tan \frac{V_n}{2} = \cot\left(\frac{V_n}{2} + \frac{\pi}{2}\right)$$

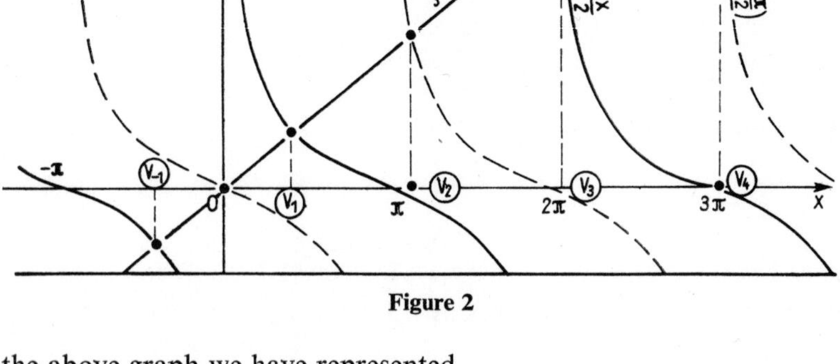

Figure 2

In the above graph we have represented,

on the one hand the curves $y = \cot \dfrac{x}{2}$ and $y = \cot\left(\dfrac{x}{2} + \dfrac{\pi}{2}\right)$, on the other $y = x$.

The solutions in V are the abscissae of the points of intersection.

Particular case $s = 2\pi$ Can it be an eigenvalue? In this case:

$$sf'' = 0 \quad \text{and} \quad f = \alpha\tau + \beta$$

Then

$$r * f = r * (\alpha\tau\beta) = \pi \int_0^1 e^{-|t-\tau|} (\alpha\tau + \beta) \, d\tau$$

$$= \pi \int_0^t e^{-t+\tau} (\alpha\tau + \beta) \, d\tau + \pi \int_t^1 e^{t-\tau} (\alpha\tau + \beta) \, d\tau$$

$$= \pi e^{-t} [e^{\tau}(\alpha\tau + \beta - \alpha)] + \pi e^{t} [e^{-\tau}(-\alpha\tau - \beta - \alpha)]_t^1$$

$$= \pi e^{-t} [e^t (\alpha t + \beta - d) - (\beta - \alpha)]$$

$$+ \pi e^{t} [(-2\alpha - \beta) e^{-1} + e^{-t} (\alpha t + \beta + \alpha)]$$

$$= \pi[e^{t-1} (-2\alpha - \beta) - e^{-t} (\beta - \alpha) + 2(\alpha t + \beta)]$$

whereas:

$$sf = s(\alpha\tau + \beta)$$

The equality

$$r * f = sf$$

is possible only if:

$$2\alpha + \beta = 0 \quad \text{and} \quad \beta - \alpha = 0$$

which leads to $\alpha = \beta = 0$, that is $f = 0$

Hence the equality is *impossible* for $f \neq 0$. We then consider the *operator*

$$Lf - g(t) = \int_0^1 r(t - \tau) f(\tau) \, d\tau$$

There exists an infinite *spectral basis*, whose elements are the above-mentioned eigenfunctions (this follows from the extension of Theorem 8 to an infinite number of dimensions).

This basis is orthogonal from what we know about hermitian operators; we can indeed verify that

$$\int_0^1 f_n(\tau) f_m(\tau) \, d\tau = 0$$

for $n \neq m$.

We can take

$$f_n = (\varrho_n + 1)\, e^{t\varrho_n} + (\varrho_n - 1)\, e^{-t\varrho_n}$$

$$f_m = (\varrho_m + 1)\, e^{t\varrho_m} + (\varrho_m - 1)\, e^{-t\varrho_m}$$

where

$$e^{\varrho_n} = \varepsilon\, \frac{\varrho_n - 1}{\varrho_n + 1} \quad \text{and} \quad e^{\varrho_m} = \varepsilon\, \frac{\varrho_m - 1}{\varrho_m + 1}$$

We have:

$$I_{nm} = \int_0^1 f_n(\tau)\, \overline{f_m}\,(\tau)\, d\tau = \int_0^1 [(\varrho_n + 1)\, \overline{(\varrho_m + 1)}\, e^{t(\varrho_n - \varrho_m)}$$

$$+ (\varrho_n + 1)\, \overline{(\varrho_m - 1)}\, e^{t(\varrho_n + \varrho_m)} + (\varrho - 1)\, \overline{(\varrho_m - 1)}\, e^{t(\varrho_n - \varrho_m)}$$

$$+ (\varrho_n - 1)\, \overline{(\varrho_m + 1)}\, e^{-t(\varrho_n + \varrho_m)}]\, dt$$

since $\varrho_m = -\varrho_m$ because the ϱ are pure complex

$$I_{nm} = \frac{(1 + \varrho_n)\,(1 - \varrho_m)}{\varrho_n - \varrho_m}\,(e^{\varrho_n - \varrho_m} - 1) - \frac{(1 + \varrho_n)\,(1 + \varrho_m)}{\varrho_n + \varrho_m}\,(e^{\varrho_n + \varrho_m} - 1)$$

$$+ \frac{(1 - \varrho_n)\,(1 - \varrho_m)}{\varrho_n + \varrho_m}\,(e^{-(\varrho_n + \varrho_m)} - 1) - \frac{(1 - \varrho_n)\,(1 + \varrho_m)}{\varrho_n - \varrho_m}\,[e^{-(\varrho_n - \varrho_m)} - 1]$$

replacing e^{ϱ_n} by $\dfrac{\varrho_n - 1}{\varrho_n + 1}$ and e^{ϱ_m} by $\dfrac{\varrho_m - 1}{\varrho_m + 1}$ we see that $I_{nm} = 0$.

On the other hand

$$\int_0^1 f_n(\tau)\, \overline{f_n}(\tau)\, d\tau = \int_0^1 |f_n(\tau)|^2\, d\tau = -(1 + \varrho_n)^2$$

$$\int_0^1 e^{2t\varrho_n}\, d\tau - (1 - \varrho_n)^2 \int_0^1 e^{-2t\varrho_n}\, d\tau + 2(1 - \varrho_n^2) \int_0^1 d\tau$$

$$= -(1 + \varrho_n)^2\, \frac{1}{2\varrho_n}\,(e^{2\varrho_n} - 1) + (1 - \varrho_n)^2\, \frac{1}{2\varrho_n}\,(e^{-2\varrho_n} - 1) + 2(1 - \varrho_n^2)$$

$$= \frac{1}{2\varrho_n}\,\underbrace{[-(\varrho_n - 1)^2 + (\varrho_n + 1)^2 + (\varrho_n + 1)^2 - (\varrho_n - 1)^2 + 4\varrho_n(1 - \varrho_n^2)]}_{= 8\varrho_n}$$

$$= 2(3 - \varrho_n^2)$$

If, for $f_n(t)$, we take the form:

$$f_n(t) = b_n[(\varrho_n + 1)\, e^{t\varrho_n} + (\varrho_n - 1)\, e^{-t\varrho_n}]$$

$$N(f_n(t)) = 2b_n^2\,(3 - \varrho_n^2)$$

whence in order to normalize we will take:

$$b_n = \frac{1}{\sqrt{2(3 - \varrho_n^2)}}$$

We can show that the system is *complete* (theorem of Parseval); f being an arbitrary function of t, we can expand it over the system of eigenfunctions: $f = \sum_{1}^{\infty} B_n f_n$ and the operator L applied to f gives

$$Lf = \sum_{1}^{\infty} B_n s_n f_n$$

where s_n is the eigenvalue associated with the eigenfunction f_n.

On the other hand a function $g(t)$ can be expanded in the same way in the form:

$$g = \sum_{1}^{\infty} A_n f_n$$

where the coefficients A_n are given by:

$$A_n = \int_{0}^{1} g(t) \overline{f_n(t)} \, dt$$

a) *The equation*

$$\underline{Lf = g}$$

leads to:

$$\sum_{1}^{\infty} B_n s_n f_n = \sum_{1}^{\infty} A_n f_n$$

or:

$$B_n s_n = A_n \quad \text{that is} \quad \boxed{B_n = \frac{A_n}{s_n}}$$

b) *Consider the equation*

$$\underline{f - \lambda Lf = g}$$

we wish to find

$$f = \sum_{1}^{\infty} B_n' f_n$$

Let:

$$\sum_{1}^{\infty} B_n' f_n - \lambda \sum_{1}^{\infty} B_n' s_n f_n = \sum_{1}^{\infty} A_n f_n$$

which gives

$$B_n'(1 - \lambda s_n) = A_n$$

and if

$$\lambda \neq \frac{1}{s_n} \qquad \boxed{B'_n = \frac{A_n}{1 - \lambda s_n}}$$

If $\lambda = \dfrac{1}{s_n}$, the problem is possible only if $A_n = 0$ that is

$$\int_0^1 g(t) f_n(t) \, dt = 0.$$

c) *Consider the equation $f'' + \lambda L f = g$*

We have seen earlier that the eigenfunctions f_n are solutions of the differential equation:

$$s_n f''_n = (s_n - 2\pi) f_n$$

Hence:

$$f''_n = \left(1 - \frac{2\pi}{s_n}\right) f_n$$

The equation reduces to:

$$\sum_1^\infty B_n \left(1 - \frac{2\pi}{s_n}\right) f_n + \lambda \sum_1^\infty s_n B_n f_n = \sum_1^\infty A_n f_n$$

whence

$$B_n \left(1 - \frac{2\pi}{s_n} + \lambda s_n\right) = A_n$$

If

$$\lambda \neq \frac{\dfrac{2\pi}{s_n} - 1}{s_n} \qquad \boxed{B_n = \frac{A_n}{1 - \dfrac{2\pi}{s_n} + \lambda s_n}}$$

If $\lambda = \dfrac{\dfrac{2\pi}{s_n} - 1}{s_n}$ we must have $A_n = 0$.

CHAPTER 2

General Theorems on Families
of Endomorphisms

COMMUTING FAMILIES OF ENDOMORPHISMS

We know that two operators commute when $AB = BA$.

PROPOSITION 4   If L admits a decomposition into a direct sum of eigen-subspaces, any endomorphism K which commutes with L, that is such that $KL = LK$, preserves these subspaces, and conversely any endomorphism which preserves the eigensubspaces of L commutes with L.

Proof   Let V be an eigensubspace of L and $x \in V$; since K and L commute we have $KLx = LKx$.

On the other hand $$Lx = \lambda x$$

hence $$KLx = K\lambda x = \lambda Kx$$
since λ is a scalar.

We deduce $$LKx = \lambda Kx$$

and hence Kx is also an eigenvector of L belonging to the eigenvalue λ, hence Kx is in the eigensubspace V.

Converse   We first recall that the endomorphisms L and K can be put in the form

$$L = \sum_i \alpha_i L_i \text{ and } K = \sum_j \beta_j K_j$$

where α_i, β_i belong to a commutative field of scalars, L_i is the element of the endomorphism L which is the identity transformation for all the eigen-vector subspaces of L other than V_i and which transforms V_i into itself, K_j is the analogous element of K.

Now we have

$$LK = \left(\sum_i \alpha_i L_i\right) \left(\sum_j \beta_j K_j\right) = \sum_i \sum_j \alpha_i \beta_j L_i K_j$$

$$KL = \left(\sum_j \beta_j K_j\right) \left(\sum_i \alpha_i L_i\right) = \sum_j \sum_i \beta_j \alpha_i K_j L_i$$

35

And since the eigensubspaces of L and K are the same, we have $K_i L_j = L_j K_i$ as can be shown from the properties of the L_i and K_j. And since the α_i and β_j also commute, we get $KL = LK$.

THEOREM IX   *A commuting family of hermitian endomorphisms admits a common orthogonal spectral basis, and conversely a family of endomorphisms which admit a common spectral basis is commuting.*

In C_n, there are at most n^2 linearly independent endomorphisms; indeed, any endomorphism induces in a system of coordinates a square $n \times n$ matrix, which is a linear combination $\sum_i \sum_j a_{ij} E_{ij}$, where E_{ij} is the matrix of order n which contains zeros everywhere except for the element i, j which is 1.

We thus reduce the family of endomorphisms (there might initially be an infinite number of them) to a finite family of n^2 endomorphisms; if we consider K endomorphisms $L_1, L_2, \ldots, L_K$ forming a basis for the family and let V be an eigensubspace of L_1, then L_2 which commutes with L_1 induces a hermitian operator in V and V is a direct sum of eigensubspaces of L_2; so that we can decompose the space into a direct sum of eigensubspaces common to L_1 and L_2; L_3 induces a hermitian operator in each one of them ...

Remark   The hermicity hypothesis is essential because it allows us to conclude that if the operator preserves a subspace the latter is a direct sum of characteristic subspaces.

Consider an endomorphism $A + iB$ where A and B are hermitian: the necessary and sufficient condition for $A + iB$ to be normal is that A and B commute.

THEOREM X   *A normal endomorphism has an orthogonal spectral basis and complex eigenvalues; conversely, an endomorphism which has an orthogonal spectral basis and complex eigenvalues is normal.*

SUMMARY

Complex representative matrices

A) L has no spectral basis

L admits a Jordan form J and the adjoint of L admits the Jordan form $\bar{J}$.

B) L has a skew spectral basis

L' admits the complementary of this basis as spectral basis.

C) L has an orthogonal spectral basis

If the eigenvalues are complex, $LL' = L'L$ and the operator is normal.

If the eigenvalues are real, the operator is hermitian, $L' = L$.

If the eigenvalues are pure imaginary, $L = -L'$, the operator is skew-hermitian.

We say an endomorphism is real if Lx is real for all real x. The matrix representative of L in a real basis is real. If a real endomorphism has a real eigenvalue, at least one real eigenvector corresponds to it. To a complex eigenvalue $\lambda + i\mu$ there corresponds the conjugate complex eigenvalue $\lambda - i\mu$ and the associated eigenvectors are conjugate complex. Hence if a zero operator has complex eigenvectors, it does not have a real basis. If all the eigenvalues are real, the Jordan form is real.

Real representative matrices

a) If it is symmetrical, we have a real and orthogonal spectral basis.

b) If the eigenvalues are real and distinct, we have a real skew spectral basis.

c) If the eigenvalues are complex and distinct, we have a complex spectral basis.

d) If the eigenvalues are real and non-distinct, we have a real Jordan form.

e) If the eigenvalues are complex and non-distinct, we have a complex Jordan form.

CHAPTER 3

Problems of Sturm–Liouville Type*

INTRODUCTION

These are spectral problems associated with certain systems of differential equations, with systems of limit conditions.

Consider the differential equation

$$\frac{d}{dx}\left[r(x)\frac{d}{dx}X \right] + [q(x) + \lambda p(x)]\,X = 0 \tag{1}$$

with *the limit conditions*

$$a_1 X(a) + a_2 X'(a) = 0$$

$$b_1 X(b) + b_2 X'(b) = 0$$

where (a, b) is the interval of definition of x, and $p(x)$, $q(x)$, $r(x)$ are assumed to be real and continuous in the open interval (a, b).

FIRST THEOREM (without proof)   *Under very general conditions of continuity and differentiability it can be proved that this system admits non-zero solutions for an infinite set of discrete values $\lambda_1, \ldots, \lambda_n, \ldots$ to which there correspond functions $X_1, \ldots, X_n, \ldots$*
   They are the eigenvalues and eigenfunctions of the problem.

SECOND THEOREM (without proof)   *Let $f(x)$ be a function which satisfies, on the interval (a, b), certain continuity and differentiability conditions. It can be expanded in the system of the X_n, and the expansion we obtain converges to $f(x)$:*

$$\underline{f(x) = \sum_{1}^{\infty} C_n X_n(x)}$$

The C_n are calculated from the X_n (if the latter form an orthonormal system, for the weight $p(x)$, cf. Theorem 3)

$$C_n = \int_{a}^{b} p(x)\, f(x)\, X_n(x)\, dx$$

---

* From *Methods of Mathematical Physics*, by Friedmann.

38

then, $\sum_1^\infty C_n X_n(x)$ converges to $f(x)$ by virtue of the Parseval theorem.

THIRD THEOREM *The X_n form an orthonormal system for the weight $p(x)$.*

Indeed, consider the differential equation for two distinct values of λ, λ_m and λ_n to which there correspond the functions X_m and X_n; we have:

$$X_n \ \bigg| \ \frac{d}{dx}[rX'_m] + [q + \lambda_m p]\,X_m = 0$$

$$X_m \ \bigg| \ \frac{d}{dx}[rX'_n] + [q + \lambda_n p]\,X_n = 0$$

Multiplying the first equation by X_n, the second by X_m and subtracting, we get:

$$(\lambda_m - \lambda_n)\,p\,X_m X_n = X_m \frac{d}{dx}(rX'_n) - X_n \frac{d}{dx}(rX'_m)$$

$$(\lambda_m - \lambda_n)\,p\,X_m X_n = \frac{d}{dx}[(rX'_n)\,X_m - (rX'_m)\,X_n]$$

Integrating from a to b, we get

$$(\lambda_m - \lambda_n)\int_a^b p(x)\,X_m X_n\,dx = [r(x)\,(X_m X'_n - X_n X'_m)]_a^b$$

Now, from the limit conditions, we have:

$$a_1 X_m(a) + a_2 X'_m(a) = 0 \quad a_1 X_n(a) + a_2 X'_n(a) = 0$$
$$b_1 X_m(b) + b_2 X'_m(b) = 0 \quad b_1 X_n(b) + b_2 X'_n(b) = 0$$

and hence:

$$\int_a^b p(x) X_m X_n\,dx = 0,$$

because we have assumed $\lambda_m \neq \lambda_n$.

If $r(a) = 0$, the first limit condition is no longer necessary.

If $r(b) = 0$, the second limit condition is no longer necessary.

If $r(a) = r(b)$ and if $X(a) = X(b)$ and $X'(a) = X'(b)$ (periodicity conditions), these new conditions are sufficient.

FOURTH THEOREM *If $p(x)$ does not change sign in the interval (a, b) (p, q, r are real), then the eigenvalues are real.*

Indeed, suppose that an eigenvalue is complex: $\lambda = \alpha + i\beta$; an eigenfunction $u + iv$ corresponds to it.

Substituting in the equation and separating real and complex parts, we get:

$$\frac{d}{dx}(ru') + (q + \alpha p)\,u - \beta pv = 0 \quad\left.\vphantom{\begin{matrix}a\\b\end{matrix}}\right|\; v$$

$$\frac{d}{dx}(rv') + (q + \alpha p)\,v + \beta pu = 0 \quad\left.\vphantom{\begin{matrix}a\\b\end{matrix}}\right|\; u$$

Multiplying the first equation by v, the second by u, and subtracting, we get:

$$-\beta(u^2 + v^2)\,p(x) = u\frac{d}{dx}(rv') - v\frac{d}{dx}(ru') = \frac{d}{dx}(urv' - vru')$$

whence:

$$-\beta \int_a^b (u^2 + v^2)\,p(x)\,dx = (urv' - vru')_a^b$$

Now in this case, the limit conditions allow us to write:

$$a_1 u(a) + a_2 u'(a) = 0 \quad b_1 u(b) + b_2 u'(b) = 0$$

$$a_1 v(a) + a_2 v'(a) = 0 \quad b_1 v(b) + b_2 v'(b) = 0$$

hence:

$$\beta \int_a^b (u^2 + v^2)\,p(x)\,dx = 0$$

and since $p(x)$ does not change sign in the interval (a, b), we get $\beta = 0$ and the eigenvalue is indeed real.

Application to a Bessel function

Consider the equation

$$x^2 y'' + xy' + (x^2 - n^2)y = 0. \tag{2}$$

This equation of Fuchs type has two solutions in the neighbourhood of the origin: J_n which is regular, Y_n which is irregular (logarithm multiplied by a polynomial).

Any solution of this equation is a (cylindrical) Bessel function. We suppose n is real; the solutions are $J_n(x)$ and we neglect for the time being the solutions $Y_n(x)$ irregular for $x = 0$ which are not suitable for many physical problems.

The expansion of the function $J_n(x)$ can be written:

$$J_n(x) = \frac{x^n}{2}\left[\frac{1}{\Gamma(n+1)} - \frac{\left(\frac{x}{2}\right)^2}{1!\,\Gamma(n+2)} + \cdots + \frac{(-1)^p\left(\frac{x}{2}\right)^{2p}}{p!\,\Gamma(n+p+1)} + \cdots\right]$$

This expansion was obtained after a suitable choice of the constant coefficients.

If n is an integer, we have:

$$J_{-n}(x) = (-1)^n J_n(x)$$

and for $n = 0$

$$J_0(x) = 1 - \frac{\left(\frac{x}{2}\right)^2}{(1!)^2} + \cdots + (-1)^p\frac{\left(\frac{x}{2}\right)^{2p}}{(p!)^2} + \cdots$$

$$J_0(0) = 1$$

For fractional n

$$J_{\frac{1}{2}}(x) = \sqrt{\frac{2}{\pi x}}\,\sin x \qquad J_{-\frac{1}{2}}(x) = \sqrt{\frac{2}{\pi x}}\,\cos x$$

From the expansion in series, we can also see that

$$xJ_n'(x) = nJ_n(x) - xJ_{n+1}(x) = -nJ_n(x) + xJ_{n-1}(x).$$

relations which can also be written

$$J_n'(x) = \frac{n}{x}J_n(x) - J_{n+1}(x)$$

$$J_n'(x) = -\frac{n}{x}J_n(x) + J_{n-1}(x)$$

By adding and subtracting side by side we get the equivalent relations

$$J_n'(x) = \frac{J_{n-1}(x) - J_{n+1}(x)}{2}$$

$$\frac{n}{x}J_n(x) = \frac{J_{n-1}(x) + J_{n+1}(x)}{2}$$

4*

On the other hand, we have

$$\frac{d}{dx}[x^n J_n(x)] = x^n J_n'(x) + nx^{n-1} J_n(x) = x^n \left[J_n'(x) + \frac{n}{x} J_n(x) \right]$$

hence

$$\frac{d}{dx}[x^n J_n(x)] = x^n J_{n-1}(x)$$

similarly we would find

$$\frac{d}{dx}[x^{-n} J_n(x)] = -x^{-n} J_{n+1}(x)$$

In particular we deduce the following from the preceding relations:

$$J_0'(x) = -J_1(x) = J_{-1}(x)$$

$$\int_0^x r J_0(r) dr = x J_1(x)$$

Indeed, differentiating this last expression, we get;

$$x J_0(x) = J_1(x) + x J_1'(x)$$

These recurrence relations allow us also to determine, by taking $n = \frac{1}{2}$, $J_{3/2}$ and more generally $J_{k+\frac{1}{2}}$ where k is an integer ≥ 0; we obtain linear combinations of $\sin x$ and $\cos x$ with polynomials in $\frac{1}{\sqrt{x}}$ as factors.

THEOREM (without proof) *For any real n, $J_n(x)$ admits an infinity of positive zeros.*

If, in equation (2), we replace x by λx, we obtain

$$x^2 \frac{d^2}{dx^2}[J_n(\lambda x)] + x \frac{d}{dx}[J_n(\lambda x)] + (\lambda^2 x^2 - n^2) J_n(\lambda x) = 0$$

Dividing by x, we get:

$$\frac{d}{dx}\left[x \frac{d}{dx} J_n(\lambda x) \right] + \left(\lambda^2 x - \frac{n^2}{x} \right) J_n(\lambda x) = 0 \qquad (1')$$

This equation is of form (1), with

$$r(x) = x$$
$$p(x) = x$$
$$q(x) = -\frac{n^2}{x}$$

where λ is replaced by λ^2.

Remark If we consider the Bessel functions, with x varying between 0 and $c > 0$, we have $r(0) = 0$ and we are hence in one of the particular cases.

We see on the other hand that $q(x)$ is continuous except for $x = 0$, unless n is equal to zero. For J_0, all the theorems will thus hold.

In fact, we can prove that the proven properties also apply if $p(x)$, $q(x)$, $r(x)$ have extremal discontinuities, hence in this case even if $n \neq 0$.

From the properties stated in the theorems 1, 2, 3, 4 we deduce that the solutions of this equation which satisfy $J_n(\lambda c) = 0$ on $(0, c)$ form a set of orthonormal functions for the weight $p(x) = x$ (the second limit condition is indeed satisfied if we set $b_2 = 0$).

Now, $J_n(\lambda c) = 0$ if λc is one of the values x_j, root of $J_n(x)$. The eigenvalues are hence $\lambda_j = \dfrac{x_j}{c}$, where x_j belongs to the infinite sequence of zeros of $J_m(x)$.

In fact, the functions $J_n(\lambda_j x)$ are orthogonal, provided that the functions $J_n'(\lambda_j x)$ are continuous (except perhaps for $x = 0$, but it can be proved that the property still holds for this case also).

Remark For each number c we have displayed an infinity of orthogonal systems (for all values of n).

The negative roots do not introduce any new eigenfunctions because $J_n(-\lambda_j x) = (-1)^n J_n(\lambda_j x)$.

We can generalize somewhat the limit conditions, since for the time being we only considered the case $J(\lambda c) = 0$, whereas the second condition is

$$b_1 X(b) + b_2 X'(b) = 0 \quad \text{with} \quad b_2 \neq 0.$$

Let indeed λ_j and λ_k be two eigenvalues; we have, since $J_n(\lambda_j x)$ and $J_n(\lambda_k x)$ satisfy (1′)

$$(\lambda_j^2 - \lambda_k^2) \int_0^c x J_n(\lambda_j x) J_n(\lambda_k x)\, dx = \int_0^c \left\{ J_n(\lambda_j x) \frac{d}{dx}\left[x \frac{d}{dx} J_n(\lambda_k x) \right] \right.$$
$$\left. - J_n(\lambda_k x) \frac{d}{dx}\left[x \frac{d}{dx} J_n(\lambda_j x) \right] \right\} dx$$

Now we readily verify that

$$J_n(\lambda_j x)\frac{d}{dx}\left[x\frac{d}{dx}J_n(\lambda_k x)\right] - J_n(\lambda_k x)\frac{d}{dx}\left[x\frac{d}{dx}J_n(\lambda_j x)\right]$$

$$= \frac{d}{dx}\left[xJ_n(\lambda_j x)\frac{d}{dx}J_n(\lambda_k x) - xJ_n(\lambda_k x)\frac{d}{dx}J_n(\lambda_j x)\right]dx$$

we suppose $n \geq 0$ because of the relation $J_{-n}(x) = (-1)^n J_n(x)$.

The integral in the second member reduces thus to:

$$\left[xJ_n(\lambda_j x)\frac{d}{dx}J_n(\lambda_k x) - xJ_n(\lambda_k x)\frac{d}{dx}J_n(\lambda_j x)\right]_0^c$$

$$= c\lambda_k J_n(\lambda_j c) J_n'(\lambda_k c) - c\lambda_j J_n(\lambda_k c) J_n'(\lambda_j c)$$

Since we have supposed $\lambda_j \neq \lambda_k$, we have orthogonalized eigenfunctions for the weight x, if

$$\frac{1}{\lambda_j c}\cdot\frac{J_n(\lambda_j c)}{J_n'(\lambda_j c)} = \frac{1}{\lambda_k c}\cdot\frac{J_n(\lambda_k c)}{J_n'(\lambda_k c)}$$

which can be written

$$\lambda c J_n'(\lambda c) = -h J_n(\lambda c),$$

where h is a constant which can be zero.

Using the relation between J_n', J_n, J_{n+1}, we have

$$(n + h) J_n(\lambda c) - \lambda c J_{n+1}(\lambda c) = 0$$

And the negative roots $-\lambda_j$ do not introduce any new eigenfunctions (cf. what we saw earlier). We suppose $h > 0$, and the general condition becomes

$$\frac{d}{dx}J_n(\lambda x) + h J_n(\lambda x) = 0.$$

All the conditions are put together in order that Theorem 4 applies; the λ_j^2 are hence real. On the other hand, the expression

$$(n + h) J_n(\lambda c) - \lambda c J_{n+1}(\lambda c) = 0$$

shows that $\lambda = 0$ will be an eigenvalue only if $J_n(0) = 0$, or if $n + h = 0$.

Now, if $J_n(0) = 0$, we deduce that $J_n(\lambda x) = 0$, hence $\lambda = 0$ is suitable as an eigenvalue with a non-zero eigenfunction, if and only if $n = h = 0$ because n and h are positive.

Calculation of the norm for orthogonal functions

Its value is

$$N_{n,j} = \int_0^c x[J_n(\lambda_j x)]^2 \, dx$$

If we multiply the Bessel equation by

$$2x \frac{d}{dx} [J_n(\lambda x)]$$

we get

$$\frac{d}{dx}\left[x \frac{d}{dx} J_n(\lambda x) \right]^2 + (\lambda^2 x^2 - n^2) \frac{d}{dx} [J_n(\lambda x)]^2 = 0.$$

We integrate this equation and the second term by parts:

$$\left\| \left[x \frac{d}{dx} J_n(\lambda x) \right]^2 + (\lambda^2 x^2 - n^2) J_n(\lambda x)^2 \right|_0^c - 2\lambda^2 N_{n,j} = 0$$

Since

$$x J_n'(x) = n J_n(x) - x J_{n+1}(x),$$

we get:

$$2\lambda^2 N_{n,j} = 2\lambda^2 \left[\frac{c^2}{2} (J_n(\lambda c)^2 + J_{n+1}(\lambda c)^2) - \frac{cn}{\lambda} J_n(\lambda c) J_{n+1}(\lambda c) \right]$$

1) λ_j is a root of $J_n(\lambda c) = 0$. In this case we have:

$$N_{n,j} = \frac{c^2}{2} [J_{n+1}(\lambda_j c)]^2$$

$$\sqrt{N_{n,j}} = \frac{c}{\sqrt{2}} |J_{n+1}(\lambda_j c)|$$

2) General case. Using the general relation

$$\lambda c J_n' = -h J_n$$

$$(n + h) J_n = \lambda c J_{n+1}$$

we get

$$(n + h) J_n(\lambda_j c) = \lambda_j c J_{n+1}(\lambda_j c)$$

Whence

$$N_{n,j} = \frac{\lambda_j^2 c^2 + h^2 - n^2}{2\lambda_j^2} [J_n(\lambda_j c)]^2$$

THEOREM (without proof)   *Under general conditions, any function $f(x)$ can be expanded as a series in such a system of eigenfunctions. If we suppose that*

$$\int_0^c \sqrt{x}\, f(x)\ dx\ \text{is absolutely convergent, we can show that the series con-}$$

verges everywhere to

$$\tfrac{1}{2}[f(x - 0) + f(x + 0)].$$

Application: The heat equation in an infinite cylinder

The equation of the cylinder is $r = c$. It is a conductor and its exterior temperature is maintained at 0°.

The initial temperature is $f(r)$ (a function of r alone).

At time t the temperature is $T = T(r, t)$, an unknown function.

The heat equations in cylindrical coordinates is:

$$\frac{\partial T}{\partial t} = k\left(\frac{\partial^2 T}{\partial r^2} + \frac{1}{r}\frac{\partial T}{\partial r}\right) \quad \text{where} \quad 0 \leqq r < c, t > 0, k > 0 \tag{1}$$

k is a constant which depends on the medium.

The limit conditions are:

$$T(c, t) = 0 \qquad \text{for} \quad t > 0$$

$$T(r, 0) = f(r) \qquad \text{for} \quad 0 < r < c \quad \text{and} \quad t = 0$$

We suppose $f(r)$ and $f'(r)$ are piecewise continuous in the interval $(0, c)$, and that for the discontinuities of $f(r)$ we have:

$$f(r) = \tfrac{1}{2}(f(r + 0) + f(r - 0))$$

Separating the variables:

$$T = X(r)\, Y(t)$$

Substituting in equation (1) we get:

$$X(r)\, Y'(t) = k\left[X''(r)\, Y(t) + \frac{1}{r} X'(r)\, Y(t) \right]$$

Dividing by $X(r)\, Y(t)$

$$\frac{Y'(t)}{k\, Y(t)} = \frac{1}{X(r)}\left[X''(r) + \frac{1}{r} X'(r) \right]$$

and these two sides, one a function of t, the other a function of r, can hence only be constant. Let λ^2 be this constant (without making any hypothesis for λ).

The limit conditions impose the minus sign in front of λ^2 because integrating at the origin we have:

$$L\frac{Y}{Y_0} = -k\lambda^2 t \quad \text{and} \quad Y = Y_0 e^{-k\lambda^2 t}$$

(an equation in t).

Indeed, when time increases, we have equilibrium and $+\lambda^2$ is impossible. The equation in r is:

$$X''(r) + \frac{1}{r}X'(r) + \lambda^2 X(r) = 0$$

It is a Bessel equation with $n = 0$. Only J_0 is suitable because the solution is regular for $r = 0$, which discards Y_0.

Since $T(c, 0) = 0$ we have $X(c) = 0$ which characterizes the system of eigenfunctions under consideration.

From Theorem 4, only the real λ occur and λ is positive because negative λ do not give rise to new eigenvalues.

The functions Y_0 are of type $\log r(P(r)) + P'(r)$ (Fuchs) and $Y_0(r)$ tends to infinity for $r \to 0$. The physical nature of the problem leads to a regular form, hence to the elimination of Y_0.

Thus if we denote by λ_j the jth root of $J_0(\lambda c) = 0$ we will have as solution $X(r)\, Y(t)$ functions of type $J_0(\lambda_j r)e^{-k\lambda^2 t}$

$$\boxed{X(r)\, Y(t) = J_0(\lambda_j r)e^{-k\lambda_j^2 t}}$$

The most general solution is:

$$\boxed{T(r, t) = \sum_{j=1}^{\infty} A_j J_0(\lambda_j r)e^{-k\lambda_j^2 t}}$$

Expressing the fact that for t tending to 0, $T = f(r)$, we expand

$$f(r) = \sum_{j=1}^{\infty} B_j J_0(\lambda_j r)$$

we determine B_j by multiplying by $J_0(\lambda_j r)$ and integrating from 0 to c,

$$\int_0^c rf(r)\, J_0(\lambda_j r)\, dr = B_j \frac{c^2}{2}[J_1(\lambda_j c)]^2$$

When t tends to 0, $T(r, 0)$ tends to $\sum A_j J_0(\lambda_j r)$, whence

$$T(r,t) = \frac{2}{c^2} \sum_{j=1}^{\infty} \frac{J_0(\lambda_j r)}{[J_1(\lambda_j r)]^2} e^{-k\lambda_j^2 t} \int_0^c r' f(r') J_0(\lambda_j r') \, dr'$$

The series does indeed tend to 0 when t tends to infinity. We note that the quantity $\dfrac{1}{\lambda_j [J_1(\lambda_j c)]^2}$ is bounded for all λ_j (by its asymptotic expansion).

Indeed, if λ_j is finite, this is obvious; on the other hand, if λ_j tends to infinity we will expand J_1 into asymptotic values and prove that $\dfrac{A_j}{\lambda_j}$ is bounded. Hence $\dfrac{B_j}{\lambda_j}$ is bounded because $f(r)$ and J_0 are bounded.

Hence for all t the terms in the series are bounded by $M\lambda_j e^{-k\lambda_j^2 t_0}$ for all t_0 with $M = $ constant.

On the other hand, $\lambda_{j+1} - \lambda_j$ tends to π because the asymptotic expansions of J_0 are circular functions.

Hence the series converges uniformly for all positive values of t and all the limit conditions are satisfied. The derivative of J_0 is bounded so that we could prove the same result for the derivative of the series and equation (1) is hence verified.

On the other hand if we apply Abel's rule we show that for t tending to 0 on the right, $T(r, t)$ tends to the $f(r)$ under consideration. The solution is hence unique (Fourier-Bessel Theorem).

Example of a problem of this type

A problem of radiation.

We suppose that the radiation flux is proportional to the temperature. The equation is then

$$- k \frac{\partial T}{\partial r} = ET$$

where k is the conductivity of the cylinder and E is the external conductivity.

If we denote by h the quantity $h = \dfrac{cE}{k}$, the eigenequation becomes

$$\lambda c J_0'(\lambda c) = -h J_0(\lambda c)$$

In the general formula the elements A_j are

$$A_j = \frac{2\lambda_j^2}{(\lambda_j^2 c^2 + h^2)\,[J_0(\lambda c)]^2} \int_0^c r J_0(\lambda_j r)\, f(r)\, dr$$

If $h = 0$ (phenomenon of reflexion or thermal isolation) we get

$$\lambda_i = 0 \quad \text{and} \quad A_i = \frac{2}{c^2} \int_0^c r f(r)\, dr = \text{constant}$$

INTEGRAL, DIFFERENTIAL, INTEGRO-DIFFERENTIAL, AND PARTIAL DIFFERENTIAL EQUATIONS

Reactor physics and nuclear physics*

The purpose of reactor theory is the determination of the behaviour of neutrons in diffusing, absorbing and multiplying media. But because of the very complex variations with respect to time of the neutrons and the cross-sections for these various possibilities, a precise description of the lives of the neutrons (which can be diffused with or without loss of energy, absorbed by or escaped from the system) requires simplifying hypotheses.

To set the problem into equations we will have to establish neutron balances (conservation equations).

In a volume element dV, the rate of change of density of neutrons is $\frac{\partial n}{\partial t}$, where n is the number of neutrons. $\frac{\partial n}{\partial t}$ is equal to the rate of production, less the rate of leakage and that of absorption.

In equilibrium $\frac{\partial n}{\partial t} = 0$ (steady state)

The fundamental conservation equation is called the *Boltzmann transport equation*. The unknown function is the angular distribution of the velocity vectors of the neutrons.

$n(\vec{r}, v, \vec{\Omega})$: number of neutrons at $\vec{r}$, per unit volume propagating with velocity v per unit velocity, in the direction $\vec{\Omega}$ per unit of solid angle.

The neutron flux Φ is defined by

$$\Phi(r, v) = \int\limits_{\substack{\text{all} \\ \text{directions}}} n(\vec{r}, v, \vec{\Omega}) v\, d\Omega$$

* From *Reactor Physics*, by Glasstone and Edlund.

There can be simplifications: isotropy, independence of the energy, etc.

If there is isotropy, n is independent of Ω and the Boltzmann equation becomes a classical diffusion equation. Under certain conditions we can apply Fick's law to the diffusion of monokinetic neutrons. Let $\vec{J}$ be a number of neutrons crossing per unit time a surface perpendicular to the flux.

We have

$$\vec{J} = -D_0 \overrightarrow{\text{grad}} \, n \quad \text{where} \quad D_0 = \text{diffusion coefficient}$$

$$\vec{J} = -D \overrightarrow{\text{grad}} \, \Phi \quad \text{where} \quad D = \text{diffusion coefficient for the flux (dimension of a length)}$$

Since $\Phi = nv$, we will have $D = \dfrac{D_0}{v}$.

Definition of the effective cross-sections. The description of the interaction of the neutrons with atomic nuclei is made quantitative by the introduction of the quantity σ equal to the probability of the occurrence of a particular nuclear reaction under given conditions; it is a specific possibility of this reaction for incident particles with a given energy. So we will have to consider the initial energy of the particle, the loss of energy during the reaction, the direction and the type of collision. We will for instance distinguish as different types of collision the elastic and the inelastic collisions depending on whether or not there is loss of energy, and as directions that before and that after the collision.

The rate will depend on the number of neutrons, their velocity and on the number and the type of nuclei of the medium under consideration. For instance, we can define a flux I of neutrons per cm^3 hitting perpendicularly in unit time a plate of the thickness of an atom of N_a atoms per cm^2.

If we denote by C the number of individual nuclear processes per cm^2 in the same time (for instance a capture) and if σ is the average number of individual paths per incident neutron in the flux I and per nucleus, we get:

$$\sigma = \frac{c}{I N_a} \, cm^2 \quad \text{per nucleus.}$$

The unit generally used is the barn $= 10^{-24} \, cm^2$.

The quantity $\sigma N_a = \dfrac{C}{I}$ is the fraction of incident beam which effectively has the action under consideration. It will hence be interpreted as the effective surface per cm^3, that is as the effective cross-section.

If we have N nuclei per cm³ for a lamina of given thickness, $N\sigma$ will be the macroscopic effective cross-section which we will denote by Σ. It is the inverse of a length.

We now introduce the attenuation of the flux of neutrons through a target of finite thickness (without taking diffusion into account). Let N be the number of nuclei/cm³ of the target. Consider a lamina of thickness x. We define the emerging flux I_x in terms of the incident flux I_0. $N\mathrm{d}x$ is the number of atoms in the slice of thickness $\mathrm{d}x$ (per cm²), $\sigma N\mathrm{d}x$ is the active fraction of incident neutrons, that is equal to $-\dfrac{\mathrm{d}I}{I}$.

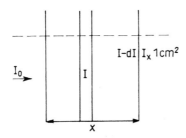

Figure 3

We have hence:

$$\sigma N\,\mathrm{d}x = -\frac{\mathrm{d}I}{I} \quad \text{whence} \quad I_x = I_0 e^{-N\sigma x} = I_0 e^{-\Sigma x}$$

If ϱ is the density of the medium in g/cm³ and if A is the atomic weight, we have:

$$N = \frac{\varrho}{A} N_0 \quad \text{where} \quad N_0 = 6.2 \times 10^{23}$$

and

$$\Sigma = \frac{\varrho N_0 \sigma}{A}.$$

If there are several types of nuclei we have

$$\Sigma = N_1\sigma_1 + N_2\sigma_2 + \cdots$$

then we replace A by the molecular mass and if ν_i is the number of nuclei of type i we will have

$$\Sigma = \frac{\varrho N_0}{M}(\nu_1\sigma_1 + \nu_2\sigma_2 + \cdots)$$

Calculation of leakages

Consider a volume dV around the point $M(r, \theta, \varphi)$.

The number of collisions will be equal to $\Sigma_1 \Phi \, dV$ where Σ_1 is the macroscopic effective scattering cross-section.

We will suppose that the phenomenon is isotropic. The probability that a neutron of volume dV be scattered in the direction dS is given by the solid angle subtended by M, namely:

$$d\Omega = \frac{ds \cos \theta}{4\pi r^2}$$

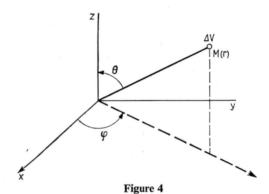

Figure 4

The probability that neutrons with direction of propagation $\vec{d\Omega}$ reach dS without any new collisions is $e^{-\Sigma r}$ where $\Sigma = \Sigma_a + \Sigma_1$, Σ_a being the macroscopic effective absorption cross-section. We suppose Σ_a is very small by comparison with Σ_1. We will then have a number of neutrons

$$\Sigma_1 \Phi \frac{ds \cos \theta}{4\pi r^2} e^{\Sigma_1 r}$$

But

$$dV = r^2 \sin \theta \, dr \, d\varphi \, d\theta$$

and the total number of neutrons scattered per second in negative z, which we will call $J_- \, ds$, will be: (neutronic current density)

$$J_- \, ds = \frac{\sum_j ds}{4\pi} \int\limits_{r=0}^{\infty} \int\limits_{\varphi=0}^{2\pi} \int\limits_{\theta=0}^{\pi/2} \Phi e^{-\Sigma_1 r} \cos \theta \sin \theta \, d\theta \, d\varphi \, dr$$

At the origin, we can expand φ into a Taylor series. We get:

$$\Phi = \Phi_0 + x\left(\frac{\partial \Phi}{\partial x}\right)_0 + y\left(\frac{\partial \Phi}{\partial y}\right)_0 + z\left(\frac{\partial \Phi}{\partial z}\right)_0 + \cdots$$

where

$$x = r \sin \theta \cos \varphi$$

$$y = r \sin \theta \sin \varphi$$

$$z = r \cos \theta$$

The terms in x and y will disappear because φ is integrated from 0 to 2π. Whence

$$J_- = \frac{\Sigma s}{4\pi}\left[\Phi_0 \iiint e^{-\Sigma_1 r} \cos \theta \sin \theta \, d\theta \, d\varphi \, dr\right.$$

$$\left. + \left(\frac{\partial \Phi}{\partial z}\right)_0 \iiint r e^{-\Sigma_1 r} \cos^2 \theta \sin \theta \, d\theta \, d\varphi \, dr\right]$$

Whence:

$$J_- = \frac{\Phi_0}{4} + \frac{1}{6\Sigma_1}\left(\frac{\partial \Phi}{\partial z}\right)_0$$

We could similarly calculate:

$$J_+ = \frac{\Phi_0}{4} - \frac{1}{6\Sigma_1}\left(\frac{\partial \Phi}{\partial z}\right)_0$$

by letting θ vary from $\pi/2$ to π. Whence

$$Jz = J_+ - J_- = -\frac{1}{3\Sigma_1}\left(\frac{\partial \Phi}{\partial z}\right)_0$$

If we had taken time into account we would have had to introduce $t = \frac{r}{v}$ and to know $\Phi\left(t - \frac{r}{v}\right)$.

Continuing the calculations of Φ to terms of the second order we see that these terms cancel for J_z. The calculation we have performed is hence valid in the second order.

On the other hand we can replace Σ_1 by $\frac{1}{\lambda_1}$ with λ the "*mean free path*".

The formula is valid for two or three mean free paths of a strong source, of an absorbant or of a boundary (because of the strong decreasing of $e^{-\Sigma_1 r}$).

Similarly we would get:

$$J_x = -\frac{\lambda_1}{3}\left(\frac{\partial\Phi}{\partial x}\right)_0$$

$$J_y = -\frac{\lambda_1}{3}\left(\frac{\partial\Phi}{\partial y}\right)_0$$

and finally

$$\vec{J} = -\frac{\lambda_1}{3}\left[\left(\frac{\partial\Phi}{\partial x}\right)_0\vec{i} + \left(\frac{\partial\Phi}{\partial y}\right)_0\vec{k} + \left(\frac{\partial\Phi}{\partial z}\right)_0\vec{l}\right]$$

where $\vec{i}, \vec{k}, \vec{l}$ are the unitary vectors on the three axes and

$$\boxed{\vec{J} = -\frac{\lambda_1}{3}\overrightarrow{\text{grad}}\,\Phi}$$

Calculation of the leakages. For this calculation we consider a volume element dV and the neutronic current which traverses this volume element in the direction of the z-axis.

There enters

$$J_z dx\,dy,$$

there goes out

$$J_{z+dz}\,dx\,dy,$$

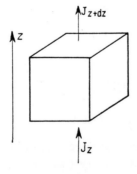

Figure 5

so that the neutronic current passing through the volume dV in the direction of the z-axis is expressed by:

$$(J_{z+dz} - J_z)\, dx\, dy = -D\left[\left(\frac{\partial \Phi}{\partial z}\right)_{z+dz} - \left(\frac{\partial \Phi}{\partial z}\right)_{z}\right] dx\, dy = -D\, \frac{\partial^2 \Phi}{\partial z^2}\, dv$$

an expression in which $D = \dfrac{\lambda_s}{3}$ is the diffusion coefficient.

Applying the same reasoning to the two other directions x and y we deduce the total number of neutrons lost by leakage during the volume dV:

$$-D\left(\frac{\partial^2 \Phi}{\partial x^2} + \frac{\partial^2 \Phi}{\partial y^2} + \frac{\partial^2 \Phi}{\partial z^2}\right) dv$$

The rate of leakage per unit volume per second is hence:

$$= -D\, \Delta\Phi$$

In fact the rate of flux of neutrons is given by the divergence of $\vec{J}$, and we get the same result:

$$\operatorname{div} \vec{J} = -D \operatorname{div} \overrightarrow{\operatorname{grad}} \Phi = -D\, \Delta\Phi$$

The variation of the number of neutrons in a volume dv in time dt is given by $\dfrac{\partial n}{\partial t}\, dv$, and this is equal to the algebraic sum of the production, the leakages and the absorption:

$$\frac{\partial n}{\partial t}\, dv = S\, dv - (-D\, \Delta\Phi)\, dv - \Sigma_a \Phi\, dv.$$

We deduce the *diffusion equation*

$$\boxed{\frac{\partial n}{\partial t} = S + D\, \Delta\Phi - \Sigma_a \Phi}$$

We carefully note the *hypotheses* which lead us to this result:

a) we considered *monokinetic neutrons*;

b) we set ourselves at *distances from the source* greater than 2 or 3 mean free paths.

On the other hand this equation is not sufficient and has no significance in itself if we do not add limit conditions. The following study will show the fundamental importance of these conditions.

CHAPTER 4

Role of Limit Conditions in the Study of Certain Operators

Suppose we characterize a physical problem by a differential, integral or integro-differential operator in an element of surface or of volume, we will also have to give limit conditions (for instance: vanishing of a function at a boundary or continuity of a derivative at a boundary, etc.). Given these limit conditions, the operator will only be their vehicle, thus propagating these conditions to the interior of the volume or the surface we are studying.

Examples* We consider the operator $L = -\dfrac{d^2}{dx^2}$.

First case—We apply L on a manifold such that

$$\boxed{u(0) = u(1) = 0}$$

(which amounts to saying that the function vanishes at the limits)

We find the eigenfunctions and the eigenvalues of L:

$$-\frac{d^2u}{dx^2} - \lambda u$$

The functions u will be of the form:

$$u = A \sin \sqrt{\lambda}\, x + B \cos \sqrt{\lambda}\, x$$

but the first limit condition requires that $B = 0$ and reduces this solution to the form:

$$\sin \sqrt{\lambda}\, x.$$

Hence the *spectral equation* will be, from the second condition:

$$\sin \sqrt{\lambda} \; = 0$$

$$\sqrt{\lambda} = n\pi$$

$$\boxed{\lambda = n^2\pi^2}, \quad \text{where} \quad n = 1, 2\cdots$$

---

* From *Methods of Mathematical Physics,* by Friedmann.

(The value $n = 0$ is excluded since then $\lambda = 0$ and the function is identically zero.)

We can find the adjoint of L, to do this we perform the product defined by:

$$\langle v, Lu \rangle = \int_0^1 - vu'' \, dx$$

$$= [- vu']_0^1 + \int_0^1 v'u' \, dx$$

$$= [-vu']_0^1 + [v'u]_0^1 - \int_0^1 uv'' \, dx.$$

Now, the adjoint operator L^* is defined by the relation:

$$\langle v, Lu \rangle = \langle L^*v, u \rangle$$

The *integral part* of the expression $\langle V, Lu \rangle$ is the *formal adjoint*.

In order that the formal adjoint be equal to the adjoint, the expression $[-vu' + v'u]_0^1$ must be zero; now

$$[- vu' + v'u]_0^1 = -v(1) \, u'(1) + v'(1) \, u(1) + v(0) \, u'(0) - v'(0) \, u(0)$$

and, from the limit conditions, the terms 2 and 4 are zero; thus there remains

$$- v(1) \, u'(1) + v(0) \, u'(0),$$

so we must have $v(0) = v(1) = 0$ in order that $L^* = - \dfrac{d^2}{dx^2}$.

If $\mathfrak{M}$ is the manifold $u(0) = u(1) = 0$, L is self-adjoint and the *eigenvalues are real* (which we have seen).

Second case — Consider, always with the same operator, the manifold for which

$$\boxed{\begin{aligned} u(0) &= 0 \\ u'(1) &= \tfrac{1}{2}u(1) \end{aligned}}$$

Again we have $\sin \sqrt{\lambda} x$ as an eigenfunction, which satisfies the equation and the first limit condition. The adjoint of L is calculated as in the first case,

$$L^* = [- vu' + v'u]_0^1 + L$$

$$= v'(1) \, u(1) - v(1) \, u'(1) + v(0) \, u'(0) - v'(0) \, u(0) + L$$

$$= v'(1) \, u(1) - v(1) \, \tfrac{1}{2}u(1) + v(0) \, u'(0) + L.$$

In order that L alone remains in the right-hand side, we must have:

$$\left\{ \begin{array}{l} v'(1) - \dfrac{v(1)}{2} = 0 \\[2mm] v(0) = 0 \end{array} \right.$$

On this manifold $\mathfrak{M}$, L is self-adjoint, the eigenvalues are all real. If the second condition is satisfied, we get:

$$\sqrt{\lambda} \, \cos \sqrt{\lambda} = \tfrac{1}{2} \sin \sqrt{\lambda}$$
$$\tan \sqrt{\lambda} = 2 \sqrt{\lambda}$$

(real λ). We can distinguish two cases:

$$\tan k = 2k$$

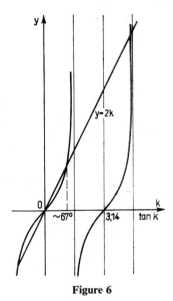

Figure 6

1) $\lambda = k^2$, thus we have $\tan k = 2k$. There are an infinity of roots which are deduced in the graph from the intersections of $y = 2k$ with $y = \tan k$;

2) $\lambda = -k^2$, whence

$$\tanh k = 2k$$

that is $k = 0$, but then $\lambda = 0$ and $u(x) = 0$, solutions which *must be discarded.*

Third case — We still consider the same operator

$$L = -\frac{d^2}{dx^2}$$

but on the manifold

$$u(0) = 0, \; u'(0) = u(1)$$

We write the product:

$$\langle v, Lu \rangle = \langle L^*v, u \rangle + [v'u - vu']_0^1,$$

now, we have:

$$[v'u - vu']_0^1 = v'(1) u(1) - v(1) u'(1) - v'(0) u(0) + v(0) u'(0)$$
$$= u(1) [v'(1) + v(0)] - v(1) u'(1)$$

If L acts on $\mathfrak{M}$, a manifold such that $u(0) = 0$, $u'(0) = u(1)$, we can define L^* on $\mathfrak{M}^*$, such that $v(1) = 0$, $v'(1) = -v(0)$.

$\mathfrak{M}^*$ is called the *dual manifold* of $\mathfrak{M}$; L is no longer self adjoint, the eigenvalues are no longer necessarily real.

If we consider $u = \sin \sqrt{\lambda} x$ as an eigenfunction, with $u'(0) = u(1)$, the eigenequation becomes:

$$\sin \sqrt{\lambda} = \sqrt{\lambda}$$

We examine the case $\lambda = \pm k^2$.

1) If we set $\lambda = + k^2$, we must have $k = 0$, the only value which satisfies the limit conditions; now $k = 0$ implies $\lambda = 0$, and there is ambiguity since $\lambda = 0$ and $u = x$ satisfy the limit conditions; to remove this ambiguity we note that, in fact, $C \sin \sqrt{\lambda}$ is a solution, where C is determined by a normalization condition, for instance $u'(0) = 1$; hence we take $C = \dfrac{1}{\sqrt{\lambda}}$ and we then consider the function:

$$\frac{\sin \sqrt{\lambda}}{\sqrt{\lambda}} x$$

which satisfies $Lu = \lambda u$ and the first limit condition; the second condition is:

$$\frac{\sin \sqrt{\lambda}}{\sqrt{\lambda}} = 1$$

whose root is $\lambda = 0$;

2) If we set $\lambda = -k^2$, the eigenvalues are *complex*; it is interesting to find the asymptotic form of these eigenvalues. Set

$$\sqrt{\lambda} = \alpha + i\beta,$$

the eigenequation becomes:

$$\sin(\alpha + i\beta) = \alpha + i\beta,$$

whence the system:

$$\begin{cases} \sin \alpha \cosh \beta = \alpha & \qquad (1) \\ \cos \alpha \sinh \beta = \beta & \qquad (2) \end{cases}$$

From (1), if α becomes very large, $\dfrac{\alpha}{\sin \alpha}$ also becomes very large so that β becomes very large and from (2) $\dfrac{\beta}{\sinh \beta} \rightarrow 0$, i.e. $\cos \alpha \rightarrow 0$, which implies the following form for α:

$$\alpha = (n + \tfrac{1}{2})\pi$$

Hence we will have:

$$\alpha_n = (n + \tfrac{1}{2})\pi + \varepsilon_n;$$

substituting this expression in (1), we get:

$$(-1)^n \cos \varepsilon_n \operatorname{ch} \beta_n = (n + \tfrac{1}{2})\pi + \varepsilon_n$$

ε_n is small, this is possible only if $n = 2m$

$$\alpha_m = (2m + \tfrac{1}{2})\pi + \varepsilon_m \qquad (3)$$

For β_m we have:

$$\cosh \beta_m = (2m + \tfrac{1}{2})\pi + \varepsilon_m$$

now

$$\cosh \beta_m = \frac{e^{\beta_m} + e^{-\beta_m}}{2} \simeq \frac{e^{\beta_m}}{2}$$

since m is large, hence

$$\beta_m = \log(4m + 1)\pi.$$

Determining the form of ε_m from (2), which can be written, using the expression (3),

$$- \sin \varepsilon_m \sinh \beta_m = \beta_m,$$

and since ε_m is small, $\sin \varepsilon_m \simeq \varepsilon_m$,

$$\varepsilon_m = - \frac{\beta_m}{\sinh \beta_m}$$

$$\simeq - \frac{2 \log(4m + 1)\pi}{(4m + 1)\pi}$$

since $\sinh \beta_m \simeq \dfrac{e^{\beta m}}{2}$ for large m.

The expression for the asymptotic expansion is then

$$\sqrt{\lambda_m} = \boxed{\alpha_m + i\beta_m = \left(2m + \frac{1}{2}\right)\pi - \frac{2 \log (4m + 1)\pi}{(4m + 1)\pi} + i \log (4m + 1)\pi}$$

In the case under consideration the eigenfunctions are not orthogonal since the operator is not hermitian; hence in order to evaluate the expression:

$$f(x) = \sum \alpha_k u_k (x)$$

we will calculate α_k by using the eigenfunctions of the adjoint $L^* = -\dfrac{d^2}{dx^2}$.

L acts on the manifold (D) such that $u(0) = 0$, $u'(0) = u(1)$
L^* acts on the manifold (D^*) such that $v(1) = 0$, $v'(1) = -v(0)$ with

$$v(x) = \frac{\sin \sqrt{\lambda}\,(1 - x)}{\sqrt{\lambda}}$$

as the eigenfunction corresponding to L^*.

$v(x)$ satisfies the second limit condition on condition that

$$-1 + \frac{\sin \sqrt{\lambda}}{\sqrt{\lambda}} = 0$$

(we find again the spectral equation). We know that $v(x) = 1 - x$ corresponds to $\lambda = 0$.

The eigenfunctions of L^* are orthogonal to the eigenfunctions of L when they correspond to different eigenvalues, i.e. if we have

$$\int_0^1 \sin \sqrt{\lambda_k}\, x \sin \sqrt{\lambda_j}(1 - x)\, dx = 0 \qquad \text{for } \lambda_j \neq \lambda_k$$

$$\int_0^1 \sin \sqrt{\lambda_k}\, x \sin \sqrt{\lambda_j}\,(1 - x)\, dx = \frac{1 - \cos \sqrt{\lambda_k}}{2} \qquad \text{for } \lambda_k = \lambda_j \neq 0$$

$$\int_0^1 x(1 - x)\, dx = \frac{1}{6} \qquad \text{for } \lambda_k = \lambda_j = 0$$

We have hence:

$$\boxed{f(x) = \alpha_0 x + \sum \alpha_k \sin \sqrt{\lambda_k}\, x}$$

where
$$\alpha_0 = 6 \int_0^1 f(x) (1 - x) \, dx,$$

$$\alpha_k = \frac{2}{1 - \cos \sqrt{\lambda_k}} \int_0^1 f(x) \, v_k(x) \, dx.$$

Fourth case–We still have $L = -\dfrac{d^2}{dx^2}$, but with

$$\boxed{u(0) = 0, \ u'(1) = \lambda u(1)}$$

i.e. the eigenvalue occurs in the limit conditions. Hence we cannot study $Lu = \lambda u$ since the domain of L depends on λ.

a) We must first extend the definition of L

We consider the spaces of vectors u with two components:
1) $u(x)$, a twice differentiable real function,
2) u_1, a real number.
And we define the *scalar product by*:

$$\langle u, v \rangle = \int_0^1 u(x) \, v(x) \, dx + u_1 v_1.$$

Consider the subspace D of the vectors u such that $u(0) = 0$, $u(1) = u_1$ and define L by means of the two-component vector

$$Lu \begin{cases} - u''(x) \\ u'(1). \end{cases}$$

The question reduces to finding u in D such that

$$Lu = \lambda u$$

and the conditions $u(0) = 0$ and $u'(1) = \lambda u(1)$ are verified.

b) To define the adjoint, consider the product:

$$\langle v, Lu \rangle = - \int_0^1 vu'' \, dx + v_1 u'(1)$$

$$= [v'u - vu']_0^1 + v_1 u'(1) - \int_0^1 v'' u \, dx$$

$$= v'(1) \, u(1) - \int_0^1 v'' u \, dx, \qquad \text{the three other terms being zero}$$

$$= \langle Lv, u \rangle$$

The adjoint of L thus reduces to the formal adjoint, since

$$\langle v, Lu \rangle = \langle L^*v, u \rangle = \langle Lv, u \rangle$$

and L defined in this way is *self-adjoint;* its eigenvalues are real and its eigenfunctions are orthogonal for the scalar product under consideration. Moreover these values are *positive,* indeed:

$$\langle u, Lu \rangle = - \int_0^1 uu'' \, dx + u(1) \, u'(1)$$

$$= [-uu']_0^1 + \int_0^1 u'^2 \, dx + u(1) \, u'(1)$$

$$= \int_0^1 u'^2 \, dx,$$

which is *positive.* The operator is positive definite, the eigenvalues are positive.

c) *Determination of the eigenvalues and eigenfunctions*

i) From the first limit condition: $u(x) \dfrac{\sin \sqrt{\lambda} \, x}{\sqrt{\lambda}}$

ii) From the second limit condition, $\cos \sqrt{\lambda} = \sqrt{\lambda} \sin \sqrt{\lambda}$ that is to say

$$\tan \sqrt{\lambda} = \frac{1}{\sqrt{\lambda}}.$$

where the eigenvalues are given by the intersections of the positive branch of the hyperbola

$$y = \frac{1}{\sqrt{\lambda}}$$

with the branches of

$$y = \tan \sqrt{\lambda}$$

Consider a vector F belonging to the space D, let u_n be the vector of D with components

$$\begin{cases} u_n(x) \\ u_n(1) \end{cases}$$

where u_n is an eigenfunction.

We will expand

$$F = \sum \alpha_n u_n(x)$$

where $\alpha_n = \dfrac{\langle F, u_n \rangle}{\langle u_n, u_n \rangle}$ in which the scalar products are those we have just defined.

We get:

$$\langle u_n, u_n \rangle = \int_0^1 \frac{\sin^2 \sqrt{\lambda_n}\, x}{\lambda_n}\, dx + \frac{\sin^2 \sqrt{\lambda}}{\lambda} = \frac{1}{\lambda_n} \left[\frac{1}{2} - \frac{\sin 2\sqrt{\lambda_n}}{4\sqrt{\lambda_n}} \right] + \frac{\sin^2 \sqrt{\lambda}}{\lambda}$$

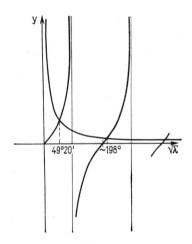

Figure 7

Hence, if F is a vector with two components, $f(x)$ and f_1, these will have the following form:

$$f(x) = \sum \alpha_n \sin \sqrt{\lambda_n}\, x$$

$$f_1 = \sum \alpha_n \frac{\sin \sqrt{\lambda_n}}{\sqrt{\lambda_n}}$$

α_n can be defined even if $f_1 \neq f(1)$, in particular if $f(x) \equiv 0, f_1 = 1$.

The four cases we have just examined for $L = -\dfrac{d^2}{dx^2}$ illustrate well the fact that for the same operator the nature of the eigenfunctions depends intimately on the limit conditions, so that the definition of the latter appears to be very important.

RETURN TO DIFFUSION EQUATIONS

First case   We consider once more the equation:

$$\frac{\partial n}{\partial t} = S + D\,\Delta\Phi - \Sigma_a\Phi$$

in order to study it in the steady case we must see that the neutronic flux is finite and non-negative in the region where we apply the diffusion equation (the limit conditions can remove certain mathematical solutions). We consider the propagation of the neutrons parallel to the x-axis and we place ourselves at the common surface of two media with different diffusion properties on which we are going to study the limit conditions (we take two planes A and B, on each side of and close to this abscissa x_0).

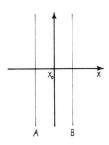

Figure 8

We get:

$$J_{A+} = J_{B+}$$
$$J_{A-} = J_{B-}$$

which implies, from the definition of J:

$$\frac{\Phi A}{4} - \frac{\lambda A}{6}\frac{\partial \Phi A}{\partial x} = \frac{\Phi B}{4} - \frac{\lambda B}{6}\frac{\partial \Phi B}{\partial x}$$

$$\frac{\Phi A}{4} + \frac{\lambda A}{6}\frac{\partial \Phi A}{\partial x} = \frac{\Phi B}{4} + \frac{\lambda B}{6}\frac{\partial \Phi B}{\partial x}$$

subtracting, then adding these two relations respectively, we get

$$-\frac{\lambda A}{3}\frac{\partial \Phi A}{\partial x} = -\frac{\lambda B}{3}\frac{\partial \Phi B}{\partial x}$$

which expresses the continuity of the current

$$\Phi A = \Phi B,$$

which expresses the continuity of the flux.

These two conditions apply to concentric spherical or coaxial cylindrical surfaces if there is isotropy.

The boundary between a diffusing medium and vacuum

In the case where the second medium is not different, there will be a finite and positive flux Φ_0 at the boundary. By extrapolation we are lead to the definition of a distance d where the flux is zero. Indeed, there is no return of the vacuum towards x_0. Hence, $J_- = 0$, i.e.

$$J_- = 0 = \frac{\Phi_0}{4} + \frac{\lambda}{6} \frac{\partial \Phi_0}{\partial x} \quad \text{for} \quad x = x_0$$

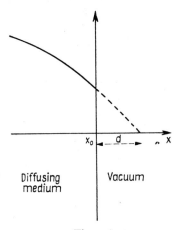

Figure 9

Since $\Phi_0 > 0$, $\dfrac{\partial \Phi_0}{\partial x} < 0$. By linear extrapolation we get:

$$-\frac{\Phi_0}{d} = \frac{\partial \Phi_0}{\partial x} = -\frac{6\Phi_0}{4\lambda_s} \qquad \boxed{d = \frac{2}{3} \lambda_s}$$

and Φ will be zero at the distance d defined in this way: it will be a limit condition: *The flux vanishes at the extrapolation distance.*

First case—The source term S is often zero except at one point, on a line or a plane. We then solve the equation outside the source region with $S = 0$ but we take S into account in the limit conditions.

In the *steady case*, we get thus $\dfrac{\partial n}{\partial t} = 0$, and, *in the absence of a source*

$(S = 0)$ the diffusion equation reduces to the classical wave equation

$$\boxed{\Delta \Phi - K^2 \Phi = 0,}$$

setting $\dfrac{\Sigma_a}{D} = K^2$ (where K^2 is the inverse of the square of a length).

Application to the case of a parallelepiped

HYPOTHESES   Consider a very long parallelepiped composed of a moderating medium and located far from the neutron source. A large fraction of the neutrons is slowed down and the flux behaves as if it originated from a plane source assumed to be located in the plane of the side $z = 0$.

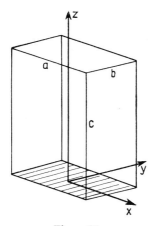

Figure 10

Let a be the dimension of the parallelepiped in the direction x, b its dimension in the direction y, c in the direction z. These dimensions take 2d (twice the extrapolated distance) into account for a and b and take d into account for c.

The diffusion equation is:

$$\Delta \Phi - K^2 \Phi = 0 \tag{1}$$

The limit conditions are:

1) non-negative finite flux;

2) $\Phi\left(\pm\dfrac{a}{2}, y, z\right) = 0,$

$\Phi\left(x, \pm\dfrac{b}{2}, z\right) = 0,$

$\Phi\left(x, y, c\right) = 0;$

3) a source condition, which we will specify later.

CALCULATIONS   We "separate" the variables and look for solutions of the form

$$\Phi = X(x)\, Y(y)\, Z(z)$$

We have:

$$\Delta\Phi = X''YZ + XY''Z + XYZ'',$$

and, substituting in equation (1) and dividing by $XYZ = \Phi$, we get:

$$\frac{I}{X}X'' + \frac{I}{Y}Y'' + \frac{I}{Z}Z'' - K^2 = 0 \qquad (1')$$

Each term is a function of only one variable and must be independent of the others.

Since K^2 is a constant, this will be possible if and only if we have:

$$\frac{I}{X}X'' = -\alpha^2$$

$$\frac{I}{Y}Y'' = -\beta^2$$

$$\frac{I}{Z}Z'' = +\gamma^2$$

where $-\alpha^2 - \beta^2 + \gamma^2 - K^2 = 0$

The signs chosen in these relations will be confirmed by the study of the limit conditions.

Indeed, if we have a form $\dfrac{I}{X}X'' = -\lambda^2$, the solution will be:

$$X = A\cos\lambda x + B\sin\lambda x$$

and if we take $\dfrac{I}{X} X'' = \lambda^2$, the solution will be:

$$= A \cosh \lambda x + B \sinh \lambda x.$$

We see immediately that in both cases the sine terms are zero if we take the axis of symmetry as the z axis, since the flux can only be represented by an even function in the direction x; moreover the flux is zero for $x = \pm \dfrac{a}{2}$, which excludes $\cosh \lambda x$ (hence $+\lambda^2$).

The solution is hence:

$$X = A \cos \alpha x,$$

and, similarly in the direction y:

$$Y = B \cos \beta y.$$

Finally γ is real, hence γ^2 is positive from (1'). Since the flux can only decrease from $z = 0$ to $z = c$ where it is zero, the solution in z is thus:

$$Z = C' \sinh \gamma(c - z)$$

which can also be set in the form:

$$Z = Ce^{-\gamma z}[1 - e^{-2\gamma(c-z)}]$$

and, for a domain which is far from both the source and the summit of the parallelepiped, we can neglect the term $e^{-2\gamma(c-z)}$.

The limit conditions $\cos \alpha \dfrac{a}{2} = 0$, $\cos \beta \dfrac{b}{2} = 0$ imply:

$$\alpha = \frac{(2m + 1)\pi}{a} \qquad X = A_m \cos \frac{(2m + 1)\pi x}{a}$$

$$\beta = \frac{(2n + 1)\pi}{b} \qquad Y = B_n \cos \frac{(2n + 1)\pi y}{b}$$

and, for any pair (m, n), γ is determined by:

$$\gamma^2_{mn} = K^2 + \left[\frac{(2m + 1)\pi}{a}\right]^2 + \left[\frac{(2n + 1)\pi}{b}\right]^2$$

and we then get the expression for Z:

$$Z = C_{mn} e^{-\gamma_{mn} z}$$

The flux can then be written as

$$\Phi = \sum_{m=1}^{\infty} \sum_{n=1}^{\infty} A_{mn} \cos \frac{(2m+1)\pi x}{a} \cos \frac{(2n+1)\pi y}{b} e^{-\gamma_{mn}z}$$

We must still determine the A_{mn}; this will be done by means of the source condition which we did not yet take into account. It is plane and constant; we can write it as follows, where δ_z is the Dirac measure:

$$S\delta_z(x, y) = \sum_{m=1}^{\infty} \sum_{n=1}^{\infty} S_{mn} \cos \frac{(2m+1)\pi x}{a} \cos \frac{(2n+1)\pi y}{b}$$

expanding it into the series of orthogonal functions satisfying the limit conditions.

We multiply by

$$\frac{(2k+1)\pi x}{a} \cos \frac{(2l+1)\pi y}{b}$$

and integrate from $-\dfrac{a}{2}$ to $+\dfrac{a}{2}$ with respect to x and from $-\dfrac{b}{2}$ to $+\dfrac{b}{2}$ with respect to y. The products such that $k = m$, $l = n$ are the only ones which do not vanish.

On the other hand, we have, by definition:

$$\int_{-\infty}^{+\infty} \int_{-\infty}^{+\infty} \delta_z(x, y)\, dx\, dy = 1$$

hence:

$$S = S_{mn} \frac{a}{2} \frac{b}{2}$$

and we deduce:

$$S_{mn} = \frac{4S}{ab}$$

We will write that the number of neutrons which leave the plane $z = 0$ per cm^2 per second for each mode (m, n) is equal to the number of neutrons emitted in this mode by the source. Since we restricted ourselves to $z > 0$, the density of the current at $z = 0$ for each mode is equal to half the number of neutrons produced in this mode:

$$J_{mn} = -D\left(\frac{\partial \Phi_{mn}}{\partial z}\right)_{z=0} = D\gamma_{mn}A_{mn}\cos\frac{(2m+1)\pi x}{a}\cos\frac{(2n+1)\pi y}{b}$$

$$= \frac{1}{2}\frac{4S}{ab}\cos\frac{(2m+1)\pi}{a}x\cos\frac{(2n+1)\pi}{b}y$$

whence we deduce:

$$A_{mn} = \frac{2S}{ab\,D\gamma_{mn}}$$

we replace A_{mn} by its value in the expression for Φ which finally becomes:

$$\Phi = \frac{2S}{abD}\sum_{m=1}^{\infty}\sum_{n=1}^{\infty}\frac{1}{\gamma_{mn}}\cos\frac{(2m+1)\pi}{a}x\cos\frac{(2n+1)\pi}{b}y e^{-\gamma_{mn}z}$$

For a pair of values m, n, the duration of the relaxation is $\dfrac{1}{\gamma_{mn}}$ (where γ_{mn} increases very rapidly with m and n).

For the first pair $m = 0$, $n = 0$ we have
$$\gamma_{00}^2 = K^2 + \frac{\pi^2}{a^2} + \frac{\pi^2}{b^2}$$

For the mode 1,0.
$$\gamma_{10}^2 = K^2 + \frac{9\pi^2}{a^2} + \frac{\pi^2}{b^2}$$

For the mode 0,1.
$$\gamma_{01}^2 = K^2 + \frac{\pi^2}{a^2} + \frac{9\pi^2}{b^2}$$

We see that the terms other than the fundamental one soon become negligible and it is this fundamental term ($m = n = 0$) only, of great importance in applications, which we need consider and we can hence take

$$\Phi \sim \frac{2S}{abD}\frac{1}{\gamma_{00}}\cos\frac{\pi x}{a}\cos\frac{\pi y}{b}e^{-\gamma_{00}z}$$

Taking the logarithmic derivative of the flux with respect to y we get:

$$\left(\frac{\partial \log \Phi(z)}{\partial z}\right)_{\substack{x=C^{te}\\y=C^{te}}} = -\gamma_{00}$$

$$x = \text{const} \quad y = \text{const}$$

We see that if we measure the logarithm of the flux we can obtain a straight line with several points in terms of z (on condition that the level points be neither too close to the source, nor too close to the end of the block under consideration), and the slope of this straight line allows us to determine γ_{00}.

We have:

$$K^2 = \gamma_{00}^2 - \left(\frac{\pi}{a}\right)^2 - \left(\frac{\pi}{b}\right)^2 = \frac{1}{L^2},$$

it is the inverse of the square of the diffusion length.

If the parallelepiped has infinite dimensions, we have:

$$K^2 = \gamma_{00}^2$$

$(\Phi(z) = Ce^{-z/L}, L = \dfrac{1}{\gamma_{00}}$, the diffusion of a plane source of thermal neutrons in an infinite medium).

Application to the case of a cylinder

HYPOTHESES   We consider two concentric cylinders with radii a and A, where the first is composed of the superposition of two cylinders characterized by the coefficients K and K' in the diffusion equation; the corresponding diffusion coefficients are respectively D and D'.

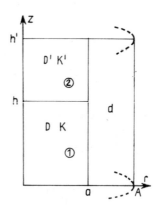

Figure 11

The cylindrical crown (a, A) is a uniquely diffusing medium with diffusion coefficient d. This system corresponds for example to the geometry we encounter in a heterogeneous pile where combustible media with different characteristics alternate, with a jacket around them.

The equations are:

$$\Delta\Phi - K^2\Phi = 0 \quad \text{for} \quad 0 < r < a \quad \text{and} \quad 0 < z < h \qquad (1)$$

$$\Delta\Phi - K'^2\Phi = 0 \qquad 0 < r < a \qquad h < z < h' \qquad (2)$$

$$\Delta\Phi + Q \quad = 0 \qquad a < r < A \qquad 0 < z < h' \qquad (3)$$

$$(Q \text{ is a constant})$$

Because of the "periodicity" or reflection at the extremities of the pile, we will have the conditions:

$$\frac{\partial\Phi}{\partial z} = 0 \quad \begin{matrix} z = 0 \\ z = h' \end{matrix} \qquad (4)$$

$$\frac{\partial\Phi}{\partial r} = 0 \quad \begin{matrix} r = 0 \\ r = A \end{matrix} \qquad (5)$$

For $0 < r < a$ and $z = h$ the flux will be continuous and there will be neutronic currents on the interface between the media (1) and (2):

$$J_h = D\left(\frac{\partial\Phi}{\partial z}\right)_{z=h^-} = D'\left(\frac{\partial\Phi}{\partial z}\right)_{z=h^+} ; \qquad (6)$$

similarly, for $0 < z < h'$ and $r = a$, at the interface of the two cylinders

$$J_a = d\left(\frac{\partial\Phi}{\partial r}\right)_{r=a^+} = \mathscr{D}\left(\frac{\partial\Phi}{\partial r}\right)_{r=a^-} \qquad (7)$$

we will have, to the left, say D, say D'

$$\mathscr{D} \begin{cases} = D & \text{for} \quad 0 < z < h \\ = D' & \text{for} \quad h < z < h' \end{cases}$$

CALCULATIONS   Consider the *first domain*, $0 < r < a$ and look for a possible expansion of Φ into a form in which the variables in r and z are separated:

$$\Phi = \sum_n a_n(r) F_n(z)$$

where we suppose that the functions $F_n(z)$ are normalized. We have:

$$\Delta\Phi = \Delta[\sum_n a_n(r) F_n(z)] = \sum_n a_n(r) \frac{\partial^2 F_n(z)}{\partial z^2} + \sum_n F_n(z) \Delta_r a_n(r)$$

a relation in which Δ_r represents the "radial" part of the Laplacian

$$\left(\Delta_r = \frac{\partial r^2}{\partial r^2} + \frac{1}{r}\frac{\partial}{\partial r}\right)$$

Hence we must have:

$$\Delta_r a_n(r) = P_n^2 a_n(r) \tag{9}$$

where P_n characterizes the eigenvalues.

This is a Bessel equation whose solutions are the functions of first and second type:

$$J_0, Y_0 \text{ if } P_n \text{ is pure complex}$$

and the modified Bessel functions of first and second type:

$$I_0, K_0 \text{ if } P_n \text{ is real.}$$

Condition (5) discards Y_0 and K_0 irregular for $r = 0$; indeed,

$$\left(\frac{\partial Y_0}{\partial r}\right)_0 = -Y_1(0) = +\infty$$

$$\left(\frac{\partial K_0}{\partial r}\right)_0 = -K_1(0) = -\infty$$

For real P_n we get

$$a_n(r) = A_n I_0(P_n r) \tag{10}$$

In the medium (1) we have:

$$\left.\begin{array}{l} f_n''^{(1)}(z) = -\alpha_n^2 f_n^{(1)}(z) \\ f_n''^{(2)}(z) = -\beta_n^2 f_n^{(2)}(z) \end{array}\right\} \tag{11}$$

with the relations:

$$-\alpha_n^2 + P_n^2 - K^2 = 0 \quad \text{and} \quad -\beta_n^2 + P_n^2 - K'^2 = 0.$$

If we consider the operator L_z such that:

$$L_z = D\left(\frac{\partial^2}{\partial z^2} - K^2\right) \quad \text{for} \quad 0 < z < h$$

$$L_z = D'\left(\frac{\partial^2}{\partial z^2} - K'^2\right) \quad \text{for} \quad h < z < h'$$

we can verify that this operator is self-adjoint

$$(v, L_z u) = \int_0^h (v)\,(L_z u)\,dz + \int_h^{h'} (v)\,(L_z u)\,dz$$

$$= \int_0^{h'} \langle L_z v, u \rangle\,dz + [D(vu' - v'u)]_0^h + [D'(vu' - v'u)]_h^{h'}$$

$$= \int_0^{h'} \langle L_z v, u \rangle\,dz$$

$$+ D[v(h)\,u'(h) - v'(h)\,u(h) - v(0)\,u'(0) + v'(0)\,u(0)]$$
$$+ D'[v(h')\,u'(h') - v'(h')\,u(h') - v(h)\,u'(h) + v'(h)\,u(h)]$$

Now u and v are zero for $z = 0$ and u' and v' are zero for $z = h'$, so that the only constant terms which remain are

$$[D(vu' - v'u)]_{h^-} - [(D'vu' - v'u)]_{h^+}$$

On the other hand the conditions (6) allow us to write

$$v^- = v^+$$
$$u^- = u^+$$
$$Du'_- = D'u'_+$$
$$Dv'_- = D'v'_+.$$

We see that these terms disappear, hence the operator is self-adjoint and the eigenvalues are exclusively real.

$$\alpha_n^2 \quad \text{and} \quad \beta_n^2 \text{ are real}$$

Suppose without loss of generality that $K' > K$.
From the conditions

$$-\alpha_n^2 + P_n^2 - K^2 = 0$$
$$-\beta_n^2 + P_n^2 - K'^2 = 0$$

we deduce

$$\boxed{\beta_n^2 = \alpha_n^2 - (K'^2 - K^2)} \tag{12}$$

and from the condition (4)

$$\left.\begin{array}{l} f_n^{(1)}(z) = u_n \cos(\alpha_n z) \\ f_n^{(2)}(z) = v_n \cos\beta_n(z - h') \end{array}\right\} \tag{13}$$

The continuity conditions of the fluxes for $z = h$ imply:

$$u_n \cos (\alpha_n h) = v_n \cos [\beta_n(h - h')]$$

so that we can take (we have not yet normalized)

$$\left. \begin{aligned} u_n &= \cos \beta_n(h' - h) \\ v_n &= \cos \alpha_n h \end{aligned} \right\} \tag{14}$$

The condition of continuity of the neutronic currents for $z = h$, $0 < r < a$, must still be satisfied,

$$\alpha_n D \cos \beta_n(h' - h) \sin (\alpha_n h) + \beta_n D' \cos \alpha_n h \sin \beta_n(h' - h) = 0$$

that is, dividing by $\cos \beta_n(h' - h) \cos \alpha_n h$

$$D\alpha_n \tan (\alpha_n h) + D'\beta_n \tan [\beta_n(h' - h)] = 0 \tag{15}$$

To obtain the *eigenvalues*, we must hence *solve the system of equations (12) and (15)*

First case—α and β are simultaneously pure complex. We set

$$\alpha_n = i\alpha_n^*$$

$$\beta_n = i\beta_n^*$$

in which α_n^* and β_n^* are both real.
We then get:

$$\frac{D\alpha_n^*}{D'\beta_n^*} = -\frac{\tanh \beta_n^* (h' - h)}{\tanh \alpha_n^* h} = \frac{1 - e^{2\beta_n^*(h'-h)}}{1 + e^{2\beta_n^*(h'-h)}} \times \frac{1 + e^{2\alpha_n^* h}}{e^{2\alpha_n^* h} - 1}$$

which is impossible by examining the possible signs for the two members. *α and β cannot be simultaneously pure complex.*

Second case—α_n is pure complex and β_n is real.

$$\alpha_n = i\alpha_n^*, \quad \text{where} \quad \alpha_n^* \text{ is real}.$$

In this case condition (12) is written:

$$\beta_n^2 + \alpha_n^{*2} = K^2 - K'^2$$

where the left side is the sum of the square two real elements, and the right side is negative since we have supposed $K' > K$. *This is hence impossible.*

Third case—α and β are simultaneously real. We will see that there are an infinity of solutions for the system (12)–(15).

Fourth case—α is real and β is pure complex. We set:

$$\beta = i\beta_n^* \quad \text{where} \quad \beta_n^* \quad \text{is real.}$$

In this case, (12) becomes:

$$\alpha_n^2 + \beta_n^{*2} = K'^2 - K^2 = \varrho^2$$

we set:

$$\alpha_n = \varrho \sin \varphi$$

$$\beta_n^* = \varrho \cos \varphi$$

and, substituting in (15), we get:

$$\frac{D}{D'} \tan \varphi \tan (h\varrho \sin \varphi) = \tanh [(h' - h) \varrho \cos \varphi] \tag{16}$$

This equation still holds if we replace φ by $\varphi + 2\pi$, φ by $-\varphi$, and φ by $\pi - \varphi$; it suffices hence to consider an interval $\left(0, \dfrac{\pi}{2}\right)$ in order to study it.

If we represent the two elements graphically we get for the first a curve which leaves 0 tangentially to 0φ and has asymptote $\varphi = \dfrac{\pi}{2}$; for the second, a curve which leaves $\tanh \varrho (h' - h)$ with a horizontal tangent at this point since the derivative $\dfrac{-\varrho(h' - h) \sin \varphi}{\cosh^2[(h' - h) \varrho \cos \varphi]}$ is zero. This curve then decreases to 0 for $\varphi = \dfrac{\pi}{2}$ and the slope is then $-\varrho(h' - h)$.

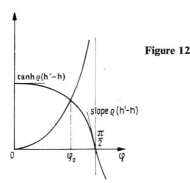

Figure 12

$\diagdown$ α β	pure complex	real
pure complex	▨	α_0 β_0^*
real	▨	infinity of solutions

Figure 13

Hence we have a solution $\alpha_0\beta_0^*$ (we can, by analogy, say that this solution is related to the connected states of quantum mechanics). We must take this carefully into account in the calculations and when establishing the formulas.

In short, the table in Figure 13 gives the possibilities.

Remark   We have associated the index zero to the particular eigenvalue we encounter; for the others we will let n vary from 1 to infinity eigenvalues of (15)]. The α_n are then all real and increase from $n = 0$. (It is easy to localize the roots of (15) when α_n and β_n are real and to show that there is a simple infinity of them.)

The equations (13) and (14) definitively characterize $f_n(z)$, either in the domain (I), or in the domain (II).

The flux is thus represented there by:

$$\Phi = \sum_{n=0}^{\infty} \frac{A_n}{N_n} I_0(p_n r) f_n(z) \tag{20}$$

where the norm N_n is defined by:

$$N_n^2 = \int_0^h [f_n^{(1)}(z)]^2 \, dz + \int_h^{h'} [f_n^{(2)}(z)]^2 \, dz \tag{21}$$

Taking the relations (12) and (15) into account, we get:

$$N_n^2 = \left\{ \frac{D'(h' - h)}{4D} [1 + \cos 2\alpha_n h] \right\} + \left\{ \frac{h}{4} [1 + \cos 2\beta_n(h' - h)] \right\}$$
$$\left\{ 1 + \frac{\sin^2 \alpha_n h}{2\alpha_n h \left[1 - \dfrac{\alpha_n^2}{K'^2 - K^2} \right]} \right\}$$

We have used the relations (12) and (13) in order to make the terms in β_n and β_n^2 vanish since the α are always real (cf. α_0) and only β is pure complex (for the first eigenvalue).

In the preceding formula it thus suffices to change $\cos 2\beta_n(h' - h)$ into $\cosh 2\beta_0(h' - h)$ to get the norm associated with the eigenvalue.

Calculation of the flux in the jacket $(a \leq r \leq A)$

We study the flux in the domain (III):

$$a < r < A$$

$$0 < z < h'$$

The diffusion equation is there:

$$\Delta\Phi + Q = 0$$

We look for a solution of the form:

$$\Phi = \Phi_1(r) + \Phi_2(r, z)$$

where

Φ_1 satisfies the equation $\Delta_r\Phi_1 + Q = 0$

Φ_2 satisfies the equation $\Delta\Phi_2 = 0$

We solve the first of these two equations which is written:

$$\left[\frac{d^2}{dr^2} + \frac{1}{r}\frac{d}{dr}\right]\Phi_1(r) + Q = 0$$

Taking the limit condition for $r = A$ into account,

$$\left[\frac{\partial\Phi}{\partial r}\right]_{r=A} = 0,$$

we have:

$$\Phi_1(r) = -\frac{Qr^2}{4} + \frac{QA^2 \log(r)}{2} + B_0 \tag{17}$$

We must still find $\Phi_2(r, z)$.

We look for the solution of Φ_2 in the form $\Sigma a_m(r) F_m(z)$, $a_m(r)$ is a solution of the equation $\Delta_r a_m(r) = q_m^2 a_m(r)$, whose two solutions are respectively $I_0(q_m r)$, regular at the origin, and $K_0(q_m r)$, irregular at the origin. We will hence have here

$$a_m(r) = \lambda_m I_0(q_m r) + \mu_m K_0(q_m r). \tag{18}$$

Condition (5) can be written:

$$q_m\lambda_m I_1(q_m A) - \mu_m q_m K_1(q_m A) = 0$$

λ_m and μ_m can then be taken respectively proportional to $K_1(q_m A)$ and $I_1(q_m A)$, whence:

$$a_m(r) = B_m[K_1(q_m A) I_0(q_m r) + I_1(q_m A) K_0(q_m r)].$$

For the calculation of $f_m(z)$, we have the equation:

$$\frac{d^2 f_m(z)}{dz^2} = q_m^2 f_m(z)$$

whence:

$$f_m(z) = C'' \cos (q_m z) + D'' \sin (q_m z).$$

For $z = 0$ and $z = h'$, we must have $\dfrac{\partial f}{\partial z} = 0$, which implies $D'' = 0$, and $q_m \sin (q_m h') = 0$, that is

$$\boxed{q_m = \frac{m\pi}{h'}}$$

Whence the required functions $f_m(z)$:

$$f_m(z) = C'' \cos (q_m z). \tag{19}$$

If we set

$$N_m^2 = \int_0^{h'} \cos^2 (q_m z) \, dz = \frac{h'}{2}$$

we determine C'' by setting

$$\int_0^{h'} f_m^2(z) \, dz = 1, \quad C''^2 \frac{h'}{2} = 1, \quad C'' = \sqrt{\frac{2}{h'}}$$

So that we finally get:

$$\boxed{\begin{aligned} \Phi = \sum_{m=1}^{\infty} \frac{B_m}{N_m} &[K_1(q_m A) I_0(q_m r) + I_1(q_m A) K_0(q_m r)] \cos \left(\frac{m\pi z}{h'}\right) \\ &+ \left[-\frac{Q r^2}{4} + \frac{Q A^2}{2} \log (r) + B_0\right] \quad a < r < A \end{aligned}} \tag{22}$$

For simplicity, we set

$$\varphi_m(r) = K_1(q_m A) I_0(q_m r) + I_1(q_m A) K_0(q_m r) \tag{23}$$

$$\psi_m(r) = K_1(q_m A) I_1(q_m r) - I_1(q_m A) K_1(q_m r) \tag{24}$$

Transition conditions (for $r = a$)

Since the unknowns must still be determined, the A_n and the B_m will be determined by the conditions expressing the continuity of the currents and of the fluxes for $r = a$.

1) *Continuity of the fluxes* We expand the $F_n(z) = \dfrac{f_n(z)}{N_n}$ in the system of the f_m:

$$\frac{f_n(z)}{N_n} = \sigma_{n0} + \sum_{m=1}^{\infty} \frac{\sigma_{nm}}{N_m} \cos (q_m z) \tag{25}$$

Whence

$$\Phi_{r=a-} = \sum_n A_n I_0(p_n a) \left\{ \sigma_{n0} + \sum_{m=1}^{\infty} \frac{\sigma_{nm}}{N_m} \cos q_m z \right\}$$

which, comparing with (22) for $r = a$, gives us, because of the continuity of the flux at $r = a$, identifying the coefficients of the constant terms and those of the terms in $\cos (q_m z)$,

$$
\text{I} \quad \left|
\begin{array}{l}
(1) \quad B_0 - \dfrac{Qa^2}{4} + \dfrac{QA^2}{2} \log (a) = \sum_n A_n \sigma_{n0} I_0(p_n a) \\[2ex]
(2) \quad B_m \varphi_m(a) = \sum_n A_n \sigma_{nm} I_0(p_n a)
\end{array}
\right.
$$

2) *Continuity of the currents* We have:

$$J_a = d \left(\frac{\partial \Phi}{\partial r} \right)_{r=a+} = \mathscr{D} \left(\frac{\partial \Phi}{\partial r} \right)_{r=a-}$$

which we can write

$$\frac{D^-}{D^+} \left(\frac{\partial \Phi}{\partial r} \right)_{r=a-} = \left(\frac{\partial \Phi}{\partial r} \right)_{r=a+}$$

with

$$\frac{D^-}{D^+} = \left\{
\begin{array}{ll}
\dfrac{D}{d} & \text{if } 0 < z < h \\[2ex]
\dfrac{D'}{d} & \text{if } h < z < h'
\end{array}
\right.$$

We expand hence

$$\frac{D^-}{D^+} \left(\frac{\partial \Phi}{\partial r} \right)_{r=a-} = \frac{C_0}{N_0} + \sum_{m=1}^{\infty} \frac{C_m}{N_m} \cos q_m z$$

We then decompose

$$\sigma_{nm} = \sigma_{nm}^{(1)} + \sigma_{nm}^{(2)} \tag{27}$$

σ_{nm} into

$$\sigma_{n0} = \sigma_{n0}^{(1)} + \sigma_{n0}^{(2)} \tag{28}$$

The index 1 is given to the integrals with respect to z taken between 0 and h; the index 2 is given to the integrals with respect to z taken between h

and h'. We have, in the domains I and II:

$$\frac{D^-}{D^+}\Phi(r) = \sum_n (A_n I_0(p_n r))\left[\frac{D}{d}\sigma_{n0}^{(1)} + \frac{D'}{d}\sigma_{n0}^{(2)}\right.$$

$$\left. + \sum_m \frac{\dfrac{D}{d}\sigma_{nm}^{(1)} + \dfrac{D'}{d}\sigma_{nm}^{(2)}}{N_m}\cos q_m z\right]$$

$$\frac{D^-}{D^+}\left(\frac{\partial\Phi}{\partial r}\right)_{r=a^-} = \sum_n (A_n p_n I_1(p_n a))\left[\frac{D}{d}\sigma_{n0}^{(1)} + \frac{D'}{d}\sigma_{n0}^{(2)}\right.$$

$$\left. + \sum_m \frac{\dfrac{D}{d}\sigma_{nm}^{(1)} + \dfrac{D'}{d}\sigma_{nm}^{(2)}}{N_m}\cos q_m z\right]$$

in the domain III:

$$\Phi = \sum_m \frac{B_m}{N_m}\varphi_m(r)\cos q_m z + \left(-\frac{Qr^2}{4} + \frac{QA^2}{2}\log r + B_0\right)$$

$$\left(\frac{\partial\Phi}{\partial r}\right)_{r=a^+} = \sum_m \frac{B_m}{N_m}q_m\psi_m(a) - \frac{Qa}{2} + \frac{QA^2}{2a}$$

Hence we have finally:

Identifying the constant terms and the terms in $\cos(q_m z)$, the system:

II
$$\begin{vmatrix} (1) & \sum_n A_n p_n I_1(p_n a)\left\{\dfrac{D}{d}\sigma_{n0}^{(1)} + \dfrac{D'}{d}\sigma_{n0}^{(2)}\right\} = \dfrac{QA^2}{2a} - \dfrac{Qa}{2} \\[2em] (2) & \sum_n A_n p_n I_1(p_n a)\left\{\dfrac{D}{d}\sigma_{nm}^{(1)} + \dfrac{D'}{d}\sigma_{nm}^{(2)}\right\} = B_m q_m \psi_m(a) \end{vmatrix}$$

The set of the systems I and II must be solved for $n = m = 0, 1, 2, \ldots$

From I(1), we get B_0 which occurs nowhere else.

From I(2), we get B_m, which we substitute in the system II which becomes

II'
$$\begin{cases} (1) & \text{unchanged} \\[1em] (2) & \sum_n A_n\left\{(\sigma_{nm}^{(1)} + \sigma_{nm}^{(2)}I_0(p_n a)\, q_m\psi_m(a) - p_n I_1(p_n a) \times \varphi_m(a)\right. \\[1.5em] & \left.\left(\dfrac{D}{d}\sigma_{nm}^{(1)} + \dfrac{D'}{d}\sigma_{nm}^{(2)}\right)\right\} = 0 \end{cases}$$

If we set

$$\varphi_0(a) = I$$

$$\psi_0(a) = 0$$

and if we make $m = 0$ in II'(2), we find again the left-hand side of II'(1); we can hence replace the system II' by

$$\sum_n A_n \left[-p_n I_1(p_n a) \, \varphi_m(a) \left\{ \frac{D}{d} \sigma_{nm}^{(1)} + \frac{D'}{d} \sigma_{nm}^{(2)} \right\} \right.$$

$$\left. + (\sigma_{nm}^{(1)} + \sigma_{nm}^{(2)}) \, q_m I_0(p_n a) \, \psi_m(a) \right] = \frac{QA^2}{2a} - \frac{Qa}{2} \delta_{m0}$$

with

$$\delta_{m0} = \begin{cases} 0 & \text{if} \quad m \neq 0 \\ 1 & \text{if} \quad m = 0 \end{cases}$$

We also set

$$A_n = \left(\frac{Qa}{2} - \frac{QA^2}{2a} \right) A^*$$

The equation above is then written:

$$\sum_n A_n^* \left[(\sigma_{nm}^{(1)} + \sigma_{nm}^{(2)}) \, I_0(p_n a) \, q_m \psi_m(a) - p_n I_1(p_n a) \, \varphi_m(a) \right.$$

$$\left. \times \left(\frac{D}{d} \sigma_{nm}^{(1)} + \frac{D'}{d} \sigma_{nm}^{(2)} \right) \right] = \delta_{m0}$$

This equation is equivalent to the linear system:

$$1 = A_1^* u_{01} + A_2^* u_{02} + \cdots + A_p^* u_{0p} + \cdots + A_n^* u_{0n} + \cdots$$

$$0 = A_1^* u_{11} + A_2^* u_{12} + \cdots + A_p^* u_{1p} + \cdots + A_n^* u_{1n} + \cdots$$

$$\cdots \cdots \cdots \cdots \cdots \cdots \cdots \cdots \cdots \cdots \cdots$$

$$0 = A_1^* u_{m1} + A_2^* u_{m2} + \cdots + A_p^* u_{mp} + \cdots + A_n^* u_{mn} + \cdots$$

The unknowns are the A_p^*, the coefficients are the u_{mn},

$$u_{mn} = \left\{ (\sigma_{nm}^{(1)} + \sigma_{nm}^{(2)}) \, I_0(p_n a) \, q_m \psi_m(a) - p_n I_1(p_n a) \, \varphi_m(a) \right.$$

$$\left. \times \left(\frac{D}{d} \sigma_{nm}^{(1)} + \frac{D'}{d} \sigma_{nm}^{(2)} \right) \right\}$$

We have hence:

$$A_p^* = (-1)^p \frac{U_{0p}}{\Delta} \tag{29}$$

with U_{0p}, the minor of the element u_{0p}.

Δ, the determinant of the coefficients.

Thus, the problem is completely solved by the determination of the A_n and B_m.

Remarks

a) Calculation of the σ_{mn}

We have:

$$f_n(z) = \begin{cases} \cos \beta_n(h' - h) \cos \alpha_n z & 0 < z < h \\ \cos \alpha_n h \cos \beta_n(z' - h) & h < z' < h' \end{cases}$$

We expand this function on the space of the $\cos \dfrac{(m\pi z)}{h'}$, whose fundamental period is $\dfrac{2\pi h'}{\pi} = 2h'$

$$f_n(z) = \sigma_{n0} + \sum_{m=1}^{\infty} \cos\left(\frac{m\pi z}{h'}\right)$$

$$\sigma_{mn} = \frac{4}{2h'} \int_0^{h'} \frac{f_n(z)}{N_n} \cos \frac{m\pi z}{h'} \, dz - \frac{2}{h'N_n} \int_0^{h} \cos \beta_n(h' - h)$$

$$\times \cos \alpha_n z \cos \frac{m\pi z}{h'} \, dz + \frac{2}{h'N_n} \int_h^{h'} \cos \alpha_n h \times \cos \beta_n(z - h') \cos \frac{m\pi z}{h'} \, dz$$

$$= \frac{\cos \beta_n(h' - h)}{h'N_n} \int_0^{h} \left[\cos\left(\alpha_n + \frac{m\pi}{h'}\right)z + \cos\left(\alpha_n - \frac{m\pi}{h'}\right)z \right] dz$$

$$+ \frac{\cos \alpha_n h}{h'N_n} \int_h^{h'} \left\{ \cos\left[\beta_n(z - h') + \frac{m\pi z}{h'}\right] \right.$$

$$\left. + \cos\left[\beta_n(z - h') - \frac{m\pi z}{h'}\right] \right\} dz$$

$$= \frac{\cos \beta_n(h' - h)}{h' N_n} \underbrace{\left[\frac{\sin\left(\alpha_n + \dfrac{m\pi}{h'}\right)h}{\alpha_n + \dfrac{m\pi}{h'}} + \frac{\sin\left(\alpha_n - \dfrac{m\pi}{h'}\right)h}{\alpha_n - \dfrac{m\pi}{h'}} \right]}_{\sigma_{nm}^{(1)}}$$

$$+ \frac{\cos \alpha_n h}{N_n h'} \underbrace{\left[-\frac{\sin\left[\beta_n(h - h') + m\pi \dfrac{h}{h'} \right]}{\beta_n + \dfrac{m\pi}{h'}} - \frac{\sin\left[\beta_n(h - h') - m\pi \dfrac{h}{h'} \right]}{\beta_n - \dfrac{m\pi}{h'}} \right]}_{\sigma_{nm}^{(2)}}$$

$$\sigma_{nm}^{(1)} = \frac{\cos \beta_n(h' - h)}{N_n h'}$$

$$\times \left[\frac{\left(\alpha_n - \dfrac{m\pi}{h'}\right)\sin\left(\alpha_n + \dfrac{m\pi}{h'}\right)h + \left(\alpha_n + \dfrac{m\pi}{h'}\right)\sin\left(\alpha_n - \dfrac{m\pi}{h'}\right)h}{\alpha_n^2 - \dfrac{m^2\pi^2}{h'^2}} \right]$$

Thus for $n, m \neq 0$

III $\begin{cases} \sigma_{nm}^{(1)} = \dfrac{\cos \beta_n(h' - h)}{N_n} \dfrac{2}{h'} \dfrac{1}{\alpha_n^2 - q_m^2} \\[2mm] \qquad \times [\alpha_n \sin \alpha_n h \cos q_m h - q_m \sin q_m h \cos \alpha_n h] \\[2mm] \text{and similarly:} \\[2mm] \sigma_{nm}^{(2)} = -\dfrac{\cos \beta_n(h - h')}{N_n} \dfrac{2}{h'} \dfrac{1}{q_m^2 - \beta_n^2} \\[2mm] \qquad \times \left[q_m \sin q_m h \cos \alpha_n h - \dfrac{D}{D'} \alpha_n \cos q_m h \sin \alpha_n h \right] \end{cases}$

For $\begin{cases} n \neq 0, \\ m = 0, \end{cases}$ take III, change $\cos \beta_n(h - h')$ into $\cosh \beta_0(h - h')$

$\qquad\qquad N_n$ into N_0

$\qquad\qquad \alpha_n$ into α_0 and β_n^2 into $-\beta_0^2$

For $n \neq 0$, $m = 0$

$$
\text{IV} \quad
\begin{cases}
\sigma_{no}^{(1)} = \dfrac{2}{h' N_n \alpha_n} \cos \beta_n (h' - h) \sin \alpha_n h, \\[4mm]
\sigma_{no}^{(2)} = -\dfrac{2}{h' N_n} \cos \beta_n (h' - h) \dfrac{D \alpha_n}{D' \beta_n^2} \sin \alpha_n h
\end{cases}
$$

For $\begin{cases} n = 0 \\ m = 0 \end{cases}$, take IV, change $\cos \beta_n (h - h')$ into $\cosh \beta_0 (h - h')$

$$N_n \quad \text{into} \quad N_0$$

$$\alpha_n \quad \text{into} \quad \alpha_0$$

$$\text{and} \quad \beta_n^2 \quad \text{into} \quad -\beta_0^2$$

b) Calculation of the eigenvalues.

Let C be the quotient $\dfrac{D}{D'}$

The eigenequation is:

$$D \alpha_n \tan (\alpha_n h) + D' \beta_n \tan \beta_n (h - h') = 0$$

with

$$\beta_n^2 = \alpha_n^2 - (K'^2 - K^2)$$

Particular case:

$$h' = 2h$$

If we set:

$$\alpha_n h = x$$

$$\beta_n (h' - h) = y,$$

we get the system:

$$Cx \tan x + y \tan y = 0 \quad C > 1$$

$$y^2 = x^2 - \lambda^2 \qquad \qquad \lambda^2 = \text{const} < 1$$

The eigenvalues are given by the intersection of the two representative curves of the system:

$$y = -\frac{Cx \tan x}{\tan [(x^2 - \lambda^2)^{\frac{1}{2}}]}, \quad y = \sqrt{x^2 - \lambda^2}$$

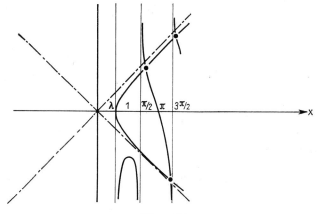

Figure 14

c) Causes of errors for the numerical computations.

Take the equation (III,I) and let $q_m = \alpha_n$. We find $\dfrac{0}{0}$. We remove the indeterminacy by setting $q_m = \alpha_n + \eta$ with small η
In the numerical calculations we will distinguish two cases:

a) $|q_m - \alpha_n| \gg \eta$: α_n will be evaluated directly from the formula III

b) $|q_m - \alpha_n| \sim \eta$: We will take a limited expansion of α_n in terms of η.

HEAT PROBLEM

We wish to evaluate the temperature distributions in the following system:
O_1, O_2 coaxial cylinders resting on a cylinder O_3
O_1 multiplying medium, radius R, height $z = \infty$,
O_2 non-multiplying medium, exterior radius S, height ∞,
O_3 non-multiplying medium, exterior radius S, height h,
ε_0 exterior medium.

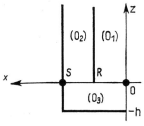

Figure 15

The physical nature of the media (2) and (3) could be different, we will suppose in the following that it is identical, without loss of generality. We denote by:

K and K' the respective thermal conductivities of O_1 and O_2 (or O_3)

Q the heat source density of O_1, assumed to be constant

β and β' the coefficients of passage respectively from O_1 to O_2 (or O_3) and from O_2 (or O_3) to ε_0,

T_0, T_1, T_2, T_3 the respective temperatures of the media ε_0, O_1, O_2 and O_3.

We recall Newton's law with respect to the temperatures T and T' of two media with conductivity K and K', with a coefficient of passage β: ($\vec{n}$ is the normal to the surface which separates the two media)

$$-K\frac{\partial T}{\partial \vec{n}} = \beta(T - T') = -K'\frac{\partial T'}{\partial \vec{n}}$$

Equations and limit conditions

for the medium O_1: $K\Delta T_1 + Q = 0$ $\qquad\qquad$ (1)

for the medium O_2: $K'\Delta T_2 = 0$ $\qquad\qquad$ (2)

for the medium O_3: $K'\Delta T_3 = 0$ $\qquad\qquad$ (3)

The condition for passing from O_1 to O_2 is written:

$$-K\frac{\partial T_1}{\partial r} = \beta(T_1 - T_2) = -K'\frac{\partial T_2}{\partial r} \quad \text{for} \quad r = R \qquad (4)$$

The condition for passing from O_2 to ε_0 is written:

$$-K'\frac{\partial T_1}{\partial r} = \beta(T_2 - T_0) \quad \text{for} \quad r = S \qquad (5)$$

The condition for passing from O_1 to O_3 is written

$$-K\frac{\partial T_1}{\partial z} = \beta(T_1 - T_3) = -K\frac{\partial T_3}{\partial z} \quad \text{for} \quad \begin{cases} z = 0 \\ 0 < r < R \end{cases} \qquad (6)$$

We must have (because of symmetry or reflexion):

$$\frac{\partial T_3}{\partial z} = 0 \quad \text{for} \quad z = -h. \qquad (7)$$

The condition for passing from O_3 to ε_0 is written:

$$-K' \frac{\partial T_3}{\partial r} = \beta'(T_3 - T_0) \quad \text{for} \quad \begin{cases} r = S \\ -h < z < 0 \end{cases} \tag{8}$$

Finally, we have the *transition* conditions *at $z = 0$* of the solutions corresponding respectively to $z > 0$ and $z < 0$, namely:

$$T_2 = T_3 \quad \text{for} \quad \begin{cases} z = 0 \\ R < r < S \end{cases} \tag{9}$$

$$\frac{\partial T_2}{\partial z} = \frac{\partial T_3}{\partial z} \quad \text{for idem.} \tag{10}$$

The functions we are looking for must also be regular for $r = 0$ and $z = \infty$. We can always suppose that $T_0 = 0$, since the solutions are of form $T_i - T_0$.

Solution of the equations for positive z.

We wish to find T in the form

$$\sum_n a_n f_n(r) g_n(z).$$

We write the Laplacian in cylindrical coordinates:

$$\Delta = \frac{\partial^2}{\partial r^2} + \frac{1}{r} \frac{\partial}{\partial r} + \frac{\partial^2}{\partial z^2} = \Delta_r + \Delta_z$$

$$\Delta f_n(r) g_n(z) = g_n(z) \Delta_r f_n(r) + f_n(r) g_n''(z)$$

We solve the homogeneous equation $\Delta T = 0$ as follows: Having separated the variables, we must have: $\dfrac{g''}{g} = -\dfrac{\Delta_r f_n}{f_n} = \text{const}$

we specify this constant:

$$\frac{g''}{g} = \pm \alpha^2$$

If we take:

$-\alpha^2$, the solution is

$$g = A \cos \alpha z + B \sin \alpha z$$

$+\alpha^2$, the solution is

$$g = A e^{\alpha z} + B e^{-\alpha z}$$

At infinity, the required solutions are independent of t and regular, which discards both the solutions periodic in z (this eliminates $-\alpha^2$) and the

7*

solutions in $e^{\alpha z}$, whence $A = 0$. We have hence:

$$\boxed{g(z) = Be^{-\alpha z}}$$

For $f_n(r)$, we then have the equation:

$$\Delta_r f_n(r) + p_n^2 f_n(r) = 0 \quad \text{where} \quad p_n^2 = \alpha^2,$$

whose solutions are:

$$J_0(p_n r) \quad \text{and} \quad Y_0(p_n r)$$

We note that in general the solution of an equation $\Delta T(r, z) = 0$ can be expressed by the table below

Constant of separation of the variables	$g(z)$	$f(r)$
positive	exponential	J_0, Y_0 at infinity $\simeq$ damped trigonometric function (cf. asymptotic expansions)
negative	trigonometric function	I_0, K_0 at infinity $\simeq$ exponential (cf. asymptotic expansions)

In the present case we must solve (1) and (2).

a) We first find the solution $u(r)$ independent of z, for infinite z because we see that physically there exist at infinity asymptotic solutions $U_1(r)$ and $U_2(r)$ for T_1 and T_2. To simplify the solution, we will first "take out" these solutions:

(1) can be written:

$$\frac{d^2 u}{dr^2} + \frac{1}{r}\frac{du}{dr} + \frac{Q}{K} = 0$$

We set

$$\frac{du}{dr} = z;$$

$$rz' + z + \frac{Q}{K}r = 0$$

without "second member":

$$rz' + z = 0 \rightarrow z = \frac{c}{r}$$

Variation of the constant:

$$z' = \frac{c'}{r} - \frac{c}{r^2} \rightarrow c' + \frac{Q}{K}r = 0$$

$$c = A - \frac{Q}{2K}r^2$$

whence

$$z = \frac{A}{r} - \frac{Q}{2K}r$$

$$u_1(r) = B + A \log r - \frac{Q}{4K}r^2$$

The regularity at the origin implies $A = 0$

$$u_1(r) = B - \frac{Q}{4K}r^2$$

similarly $\qquad u_2(r) = A + C \log r \quad$ from (2)

Determination of the constants A, B and C is done by means of the equations (4) and (5):

$$(4) \quad -K\left(-\frac{Q}{2K}R\right) = \beta\left(B - \frac{Q}{4K}R^2 - A - C \log R\right) = -K'\frac{C}{R}$$

$$(5) \quad -K'\frac{C}{S} = \beta'(A + C \log S)$$

that is

$$C = -\frac{QR^2}{2K'} \qquad\qquad A = \frac{QR^2}{2K'}\left[\log S + \frac{K'}{\beta'S}\right]$$

$$B = \frac{QR^2}{2K'}\left[\log \frac{S}{R} + \frac{K'}{\beta'S}\right] + \frac{QR^2}{4K} + \frac{QR}{2\beta}$$

b) *If z is finite*, we now have two homogeneous equations [(1) and (2)]. We hence look for T in the form

$$T - T_0 = u(r) - \Phi(r, z) \quad \text{[where we must determine } \Phi]$$

(the minus sign corresponds to physical reality which lets us foresee that the first coefficients in the expansion of Φ will now be positive).

Then:

(1) is written:

$$K\Delta T_1 + Q = K\Delta U_1 - K\Delta\Phi_1 + Q = -K\Delta\Phi_1 = 0$$

(2) is written:

$$K'\Delta T_2 = K'\Delta U_2 - K'\Delta\Phi_2 = -K'\Delta\Phi_2 = 0$$

We set:

$$\boxed{T - T_0 = u(r) - \Sigma\, A_n e^{-p_n z} F_n(r),}$$

where F_n is normalized

In O_1, the solution must be regular for $r = 0$. We have thus, since $f_n(r)$ is not normalized:

$$\text{in } O_1: \quad f_n(r) = \lambda_n J_0(p_n r)$$

$$\text{in } O_2: \quad f_n(r) = \mu_n J_0(p_n r) + \nu_n Y_0(p_n r)$$

We determine the coefficients and the eigenvalues by means of the equations (4) and (5), which give, recalling that:

$$J_0'(z) = -J_1(z) \quad \text{and} \quad Y_0'(z) = -Y_1(z)$$

(4) $$K p_n \lambda_n J_1(p_n R) = K' p_n[\mu_n J_1(p_n R) + \nu_n Y_1(p_n R)]$$

$$= \beta[\lambda_n J_0(p_n R) - \{\mu_n J_0(p_n R) + \nu_n Y_0(p_n R)\}]$$

(5) $$K' p_n[\mu_n J_1(p_n S) + \nu_n Y_1(p_n S)] = \beta'[\mu_n J_0(p_n S) + \nu_n Y_0(p_n S)]$$

We have hence a linear and homogeneous system in λ_n, μ_n and ν_n. We get a solution different from the trivial solution if the determinant Δ of the coefficients is zero; that is:

$$\begin{vmatrix} K J_1(p_n R) & -K' J_1(p_n R) & -K' Y_1(p_n R) \\ \beta J_0(p_n R) & -\beta J_0(p_n R) & -\beta Y_0(p_n R) \\ -K p_n J_1(p_n R) & & \\ 0 & \beta' J_0(p_n S) & \beta' Y_0(p_n S) \\ & -K' p_n J_1(p_n S) & -K' p_n Y_1(p_n S) \end{vmatrix} = 0$$

Which gives the equation for the eigenvalues p_n.

Orthogonality of the system

Consider the expression:

$$I = \int (f_m \Delta f_n - f_n \Delta f_m) \, r \, dr$$

where f_m and f_n are two eigenfunctions
It can be written:

$$I = \int f_m (- p_n^2 f_n) - f_n(- p_m^2 f_m) r \, dr = - \int [p_n^2 - p_m^2] f_n f_m r \, dr$$

Now

$$r \Delta f = r \frac{d^2 f}{dr^2} + \frac{df}{dr} = \left(r \frac{df}{dr} \right)'$$

whence

$$I = \int (f_m \Delta f_n - f_n \Delta f_m) \, r \, dr = \int \left[f_m \left(r \frac{df_n}{dr} \right)' - f_n \left(r \frac{df_m}{dr} \right)' \right] dr$$

Integrating by parts:

$$\int f_m (r f_n')' \, dr = r f_m f_n' - \int r f_n' f_m' \, dr$$

$$\int f_n (r f_m')' \, dr = r f_n f_m' - \int r f_m' f_n' \, dr$$

whence:

$$I = r(f_m f_n' - f_n f_m')$$

We calculate $I = \int_0^S f_n f_m \, r \, dr = 0$, taking the discontinuity for $r = R$ into account.

$$I = \int_0^S = \int_0^{R^-} + \int_{R^+}^S = I(S) - I(R^+) + I(R^-) - I(0)$$

Now:
I(0) is zero for $r = 0$, because of the factor r, f_m and f_n being regular for $r = 0$.
I(S) is zero for $r = S$, from condition (5) which is written:

$$\frac{dT_2}{dr} = - \frac{\beta'}{K} T_2 (T_0 = 0) \quad \text{that is} \quad \frac{T_2'}{T_2} = - \frac{\beta'}{K'},$$

which gives:

$$\frac{f_n'}{f_n} = \frac{f_m'}{f_m} = - \frac{\beta'}{K'}$$

for $r = R$, from condition (4)

for R^+ $-K' \dfrac{dT_2}{dr} = -\beta(T_2 - T_1)$

$\qquad\qquad\qquad\qquad\qquad\qquad\qquad$ whence $\left\{\begin{array}{l} f_m'^+ = \dfrac{K}{K'} f_m'^- \\[2mm] f_m^+ = f_m^- + \dfrac{K}{\beta} f_m'^- \end{array}\right.$

for R^- $-K \dfrac{dT_1}{dr} = \beta(T_1 - T_2)$

and

$$I(R^+) = (f_m^- f_n'^- - f_n^- f_m'^+) \frac{K}{K'} = \frac{K}{K'} I(R^-)$$

Hence

$$I = \left(1 - \frac{K}{K'}\right) I(R^-)$$

Hence if we set:

$$g_n(r) = \left\{\begin{array}{ll} K f_n(r) & \text{in } O_1 \quad 0 < r < R \\[2mm] K' f_n(r) & \text{in } O_2 \quad R < r < S \end{array}\right.$$

Then it follows from above that:

$$\int_0^S g_n(r)\, g_m(r)\, r\, dr = 0$$

This is written more explicitly:

$$K \int_0^R \lambda_n J_0(p_n r)\, \lambda_m J_0(p_m r)\, r\, dr + K' \int_R^S [\mu_n J_0(p_n r) + \nu_n Y_0(p_n r)]$$

$$\times\, [\mu_m J_0(p_m r) + \nu_m Y_0(p_m r)]\, r\, dr = 0$$

The functions $f_n(r)$ and $g_n(r)$ form a biorthogonal system which will also be normed if:

Normalization condition

$$\int_0^S f_n(r)\, g_n(r)\, r\, dr = 1$$

This condition, together with the homogeneous system above, determines λ_n, μ_n, ν_n *uniquely.*

Using the g_n allows us to calculate without difficulty the coefficients of the expansions which will be constructed.

Remark We use the following property to simplify the eigenequation:

$$Y_0(z) J_1(z) - Y_1(z) J_0(z) = \frac{2}{\pi z}$$

Indeed:

$$Y_0 J_1 - Y_1 J_0 = - Y_0 J_0' + J_0 Y_0'$$

Set:

$$W = z[- Y_0 J_0' + J_0 Y_0']$$

$$W' = - Y_0 J_0' + J_0 Y_0' + z [- Y_0 J_0'' + J_0 Y_0'']$$

$$= - Y_0 J_0' + J_0 Y_0' + z\left[- Y_0\left(-\frac{1}{z}J_0' - J_1\right) + J_0\left(-\frac{1}{z}Y_0' - Y_0\right)\right]= 0$$

$W = \text{const} \left(= \dfrac{2}{\pi}\right)$. In order to calculate the constant, we must *use the*
asymptotic expansions

$$J_0 = \sqrt{\frac{2}{\pi z}} \cos\left(z - \frac{\pi}{4}\right) \qquad Y_0 = \sqrt{\frac{2}{\pi z}} \sin\left(z - \frac{\pi}{4}\right)$$

$$J_1 = \sqrt{\frac{2}{\pi z}} \cos\left(z - \frac{3\pi}{4}\right) \qquad Y_1 = \sqrt{\frac{2}{\pi z}} \sin\left(z - \frac{3}{4}\pi\right)$$

Precisions on the eigenequation.
  It can be written:

$$0 = \frac{\dfrac{\beta'}{K'} Y_0(p_n S) - p_n Y_1(p_n S)}{\dfrac{\beta'}{K'} J_0(p_n S) - p_n J_1(p_n S)}$$

$$- \frac{\dfrac{2\beta}{\pi R K'} + p_n Y_1(p_n R)\left[\left(\dfrac{K}{K'} - 1\right)\dfrac{\beta}{K} J_0(p_n R) + p_n J_1(p_n R)\right]}{p_n J_1(p_n R)\left[\dfrac{\beta}{K}\left(\dfrac{K}{K'} - 1\right) J_0(p_n R) + p_n J_1(p_n R)\right]}$$

To calculate the roots, we must localize them, then interpolate around the corresponding values.

To facilitate the localization, we can use the asymptotic expansions of the Bessel functions.

We set:

$$\frac{\beta' S}{K'} - \frac{3}{8} = a \qquad \frac{\beta R}{K}\left(\frac{K}{K'} - 1\right) + \frac{3}{4} = b$$

$$\frac{\beta R}{K}\left(\frac{K}{K'} + 1\right) = c \qquad z = p_n S \quad \xi = p_n R$$

The asymptotic equation is written, restricting ourselves to the terms in $\dfrac{1}{p_n}$:

$$0 = \sin(z - \xi)\cos\left(\frac{\pi}{4} + z\right) - \frac{a}{z}\cos(z - \xi)\cos\left(\frac{\pi}{4} + z\right)$$

$$+ \frac{c}{2}\sin\left(z - \frac{\pi}{4} - \frac{b}{2\xi}\right)\cos\left(z - \frac{\pi}{4} - 2\xi\right)$$

If p_n is large, we need only take the first term of the equation into consideration, hence:

$$z - \xi = \varepsilon_1 + n_1\pi \quad \text{and} \quad \frac{\pi}{4} + \xi = \frac{\pi}{2} + \varepsilon_2 + n_2\pi$$

Whence the two sequences of eigenvalues:

$$p_{n_1} = \frac{\varepsilon_1 + n_1\pi}{S - R} \qquad p_{n_2} = \frac{1}{R}\left(\frac{\pi}{4} + \varepsilon_2 + n_2\pi\right)$$

Calculation of T in O_3 (solution for negative z)
    The equation we use is (3):

$$\Delta T_3 = 0$$

The solution must be regular for $r = 0$, and satisfy the limit conditions:

(7) $$\frac{\partial T_3}{\partial z} = 0 \qquad \text{for} \quad \begin{array}{l} z = -h \\ 0 < r < S \end{array}$$

(8) $$-K'\frac{\partial T_3}{\partial r} = \beta' T_3 \quad \text{for} \quad \begin{array}{l} r = S \\ -h < z < 0 \end{array}$$

We deduce, after separating the variables: (cf. the first part)

$$T_3(r, z) = \sum_{m=1}^{\infty} b_m \cosh\left[q_m(z + h)\right] J_0(q_m r)$$

If we set $q_m S = x$, the relation (8), which can be written:

$$K' q_m J_1(q_m S) = \beta' J_0(q_m S)$$

becomes

$$\frac{x J_1(x)}{J_0(x)} = \frac{\beta' S}{K'}.$$

This is the equation having eigenvalues q_m, whose zeros are readily determined since $\beta' S$ and K' are given and the function $\dfrac{x J_1(x)}{J_0(x)}$ is tabulated.

Normalization   In order that the function in r be normalized, we will modify b_m and we will replace $J_0(q_m r)$ by $\lambda_m J_0(q_m r)$ such that:

$$\int_0^S [\lambda_m J_0(q_m r)]^2 \, r \, dr = 1$$

that is:

$$\frac{1}{\lambda_m^2} = \int_0^S [J_0(q_m r)]^2 \, r \, dr.$$

T_3 is then written:

$$T_3 = \sum_{m=1}^{\infty} b_m \cosh [q_m(z + h)] \, F_m(r)$$

where b_m now denotes the value of b_m corresponding to the normalized function in r.

Transition for $z = 0$ We use the limit conditions:

a) $T_1 = T_3$ for $z = 0, 0 < r < R$.

b) $-K \dfrac{\partial T_1}{\partial z} = \beta(T_1 - T_3) = -K' \dfrac{\partial T_3}{\partial z}$ for $z = 0, 0 < r < R$

and the conditions of continuity between O_2 and O_3 which physically constitute only one medium:

c) $T_2 = T_3$ for $z = 0, R < r < S$.

d) $\dfrac{\partial T_2}{\partial z} = \dfrac{\partial T_3}{\partial z}$ for $z = 0, R < r < S$.

We can write these five equations in the following manner:

a) $T = T_3$

e) $-\dfrac{K}{B}\dfrac{\partial T}{\partial z} = T - T_3 = -\dfrac{K'}{B}\dfrac{\partial T_3}{\partial z}$

by setting:

in $(0, R)$ $K = k$ $K' = k'$ $B = \beta$ $T = T_1$

in (R, S) $K = K' = k'$ $B = \infty$ $T = T_2$

Choice of the basis system for the expansion of the solutions

In order to translate the preceding conditions, we must first expand all the functions in the same complete system. We will take that of the F_n, for, physically we can foresee, because of the better form assumed by the solutions under consideration, that the convergence will be more rapid (this form is related to the dimension of the medium, because O_3 only forms a "plug").

We will thus expand in the system of the $F_n(r)$.

1) The $U(r)$

$$U_1(r) = B - \frac{Q}{4K} r^2 \quad 0 < r < R$$

$$U_2(r) = A + C \log r \quad R < r < S$$

We set:

$$U(r) = \sum_n \alpha_n F_n(r)$$

with:

$$\alpha_n = \int_0^S U(r) \, G_n(r) \, r \, dr$$

$$= \int_0^R \left(B - \frac{Q}{4K} r^2 \right) \lambda_n J_0(p_n r) \, r \, dr + \int_R^S (A + C \log r)$$

$$\times \left[\mu_n J_0(p_n r) + \nu_n Y_0(p_n r) \right] r \, dr$$

2) The $F_m(r)$

$$F_m(r) = \sum_n \sigma_{mn} F_n(r)$$

$$\sigma_{mn} = \int_0^S F_m(r)\, G_n(r)\, r\, dr$$

$$\sigma_{mn} = \int_0^R F_m(r)\, \lambda_n J_0(p_n r)\, r\, dr + \int_R^S F_m(r)\, [\mu_n J_0(p_n r) + \nu_n Y_0(p_n r)]\, r\, dr$$

$$= \sigma_{mn}^{(1)} + \sigma_{mn}^{(2)}$$

3) We expand:

$$\frac{K'}{K}\frac{\partial T_3}{\partial z}$$

$$\frac{K'}{K}\frac{\partial T_3}{\partial z} = \sum_n P_n(z)\, F_n(r)$$

$$P_n(z) = \int_0^S \frac{K'}{K}\frac{\partial T_3}{\partial z}\, G_n(r)\, r\, dr$$

where K has the values K and K' in $(0, R)$ and (R, S) respectively and $G_n(r)$ the two respective determinations we have already used.

A) We expand:

$$\frac{K'}{B}\frac{\partial T_3}{\partial z}$$

$$\frac{K'}{B}\frac{\partial T_3}{\partial z} = \sum_n V_n(z)\, F_n(r)$$

$$V_n(z) = \int_0^S \frac{K'}{K}\frac{\partial T_3}{\partial z}\, G_n(r)\, r\, dr$$

We specify:

$$P_n(z) = \int_0^R \frac{K'}{K}\sum_m b_m q_m \sinh\left[q_m(z + h)\right] F_m(r)\, G_n(r)\, r\, dr$$

$$+ \int_R^S \sum_m b_m q_m \sinh\left[q_m(z + h)\right] F_m(r)\, G_n(r)\, r\, dr$$

and, permuting $\int$ and Σ:

$$P_n(z) = \frac{K'}{K} \sum_m b_m q_m \sinh q_m(z + h) \left[\sigma_{mn}^{(1)} + \sigma_{mn}^{(2)} \frac{K}{K'} \right]$$

$$V_n(z) = \int_0^R \frac{K'}{\beta} \sum_m b_m q_m \sinh [q_m(z + h)] F_m(r) G_n(r) r \, dr$$

$$V_n(z) = \frac{K'}{\beta} \sum_m b_m q_m \sinh [q_m(z + h)] \sigma_{mn}^{(1)}$$

We are now in position to translate the transition conditions in order to determine the unknown coefficients (A_n and b_n)

$$\Rightarrow \frac{\partial T_1}{\partial z} = \frac{K'}{K} \frac{\partial T_3}{\partial z} \quad \text{for} \quad z = 0$$

that is

$$\sum_n A_n p_n e^{-p_n z} F_n(r) = \sum_n P_n(z) F_n(r)$$

setting $z = 0$ and identifying term by term we get

$$\boxed{A_n p_n = P_n(0)}$$

$$\Rightarrow T_3 - T_1 = \frac{K'}{B} \frac{\partial T_3}{\partial z} \quad \text{for} \quad z = 0$$

that is:

$$\sum_n b_n \cosh [q_m(z + h)] \sum_n \sigma_{mn} F_n(r) - \left[\sum_n \alpha_n F_n(r) - \sum_n A_n e^{-p_n z} F_n(r) \right]$$

$$= \sum_n V_n(z) F_n(r)$$

setting $z = 0$ and identifying term by term we get

$$\boxed{-\alpha_n + A_n + \sum_m b_m \cosh (q_m h) \sigma_{mn} = V_n(0)}$$

Since $P_n(0)$ and $V_n(0)$ are known, we deduce A_n.
The second equation can be written:

$$-\alpha_n + \frac{P_n(0)}{p_n} + \sum_m b_m \cosh (q_m h) \sigma_{mn} = V_n(0)$$

$P_n(0)$ and $V_n(0)$ are expressed by series with index m, which allows us to write, setting $b_m \cosh(q_m h) = B_m$

$$\boxed{\alpha_n = \sum_m B_m u_{nm}}$$

We will solve this equation successively for $m = n = 1, 2, 3, \cdots$ *The problem is hence completely solved.*

Solution of the system $\qquad \alpha_n = \Sigma B_m u_{nm}$

$$
\begin{array}{ll}
\overset{q-1}{} & \overset{q}{} \\
U_{11}B_1 + U_{12}B_2 + U_{13}B_3 + \cdots + U_{1(q-1)}B_{q-1} \; + U_{1q}B_q & = \alpha_1 \\
U_{21}B_1 + U_{22}B_2 + U_{23}B_3 + \cdots + U_{2(q-1)}B_{q-1} \; + U_{2q}B_q & = \alpha_2
\end{array}
$$

$$. \quad . \quad . \quad . \quad . \quad . \quad . \quad . \quad . \quad . \quad . \quad . \quad . \quad . \quad . \quad . \quad . \quad . \quad . \quad . $$

$$U_{(q-1)1}B_1 + U_{(q-1)2}B_2 + U_{(q-1)3}B_3 \cdots$$

$$+ U_{(q-1)\,(q^-1)3}B_{q-1} \qquad\qquad + U_{(q-1)}B_q = \alpha_{q-1}$$

$q-1$ ------------------------------------

$$U_{q-1}B_1 + U_{q2}B_2 + U_{q3}B_3 + \cdots + U_{q(q-1)}B_{q-1} + U_q B_q \qquad = \alpha_q$$

q ------------------------------------

Let B_m^{q-1} be the solutions of the system of order $q - 1$. We set $B_m = B_m^{q-1} + B_m'$ with $B_q^{q-1} = 0$.

Subtracting term by term the homologous equations of the systems of order $(q - 1)$ and q, we get:

$$\sum_m u_{nm} B_m' = 0 \quad \text{for} \quad n = 1, 2, \cdots (q - 1)$$

and

$$\sum_m u_{qm} B_m' = \alpha_q - \sum_{m=1}^{q-1} u_{qm} B_m^{q-1} \equiv \alpha_q'$$

$$B_m' = (-1)^{m+q}\frac{U_{qm}}{\varDelta}\alpha_q'$$

U_{qm}: minor of u_{qm} in the system of order q
$\varDelta$: determinant of this system.

Suppose we have solved the system of order $(q - 1)$; for the order q it suffices to calculate the corrections B_m' of the B_m^{q-1} and to determine the new unknown B_q.

Numerically we have the following advantages:

a) the B'_m are of the same order of magnitude as α'_q and α'_q decreases as q increases.

b) the principal diagonal of the matrix of the system becomes of great importance when q increases, whence the possibility of solution by means of a relaxation method.

c) We can define simple criteria for convergence which stop the iteration when the convergence is sufficient.

Example   If we expand the function $\psi(r)$, we form the quantity:

$$\delta = \int_0^S [\psi(r) - \sum_m \eta_n F_n(r)] \, [\psi(r) - \sum_m \xi_n G_n(r)] \, r \, dr$$

$$= \int_0^S \psi^2(r) \, r \, dr - \sum_n \eta_n \xi_n$$

Testing δ, we can follow the convergence of the function (the problem is solved in the sense of minimum quadratic deviation) and we have a precise idea of the number of terms we must use.

CHAPTER 5

Perturbation Problems

We wish to solve, for $z > 0$:

$$\boxed{\Delta T + \psi(z)\,\varphi(r) = 0} \tag{1}$$

in the frame of the preceding problem of thermodynamics.

We can look for a solution with separated variables in the form:

$$T = \sum f(z)\,g(r)$$

but it is obviously in our interest to consider the preceding problem as a particular case t of the present problem, that is to say that we expand T in the form:

$$T = \sum_n [\psi_n(z) - A_n e^{-p_n z}]\, F_n(r) \tag{2}$$

introducing the "unperturbed" solution we have already studied. We get thus:

$$\Delta T = \sum_n \{[\psi_n(z) - A_n e^{-p_n z}]\,\Delta r(F_n(r)) + [\psi_n''(z) - p_n^2 A_n e^{-p_n z}]\, F_n(r)\}$$

Now

$$\Delta_r(F_n(r)) = -p_n^2 F_n(r)$$

whence:

$$\Delta T = \sum_n [\psi_n''(z) - p_n^2 \psi_n(z)]\cdot F_n(r)$$

We expand on the other hand, $\varphi(r)$ in the system of the $F_n(r)$

$$\varphi(r) = \sum_n q_n F_n(r)$$

Equation (1) is finally written:

$$\sum_n [\psi_n''(z) - p_n^2 \psi_n(z) + q_n \psi(z)]\, F_n(r) = 0$$

From the separation of the variables we can say that each coefficient of the F_n must be zero, that is:

$$\boxed{\psi_n''(z) - p_n^2 \psi_n(z) + q_n \psi(z) = 0} \tag{3}$$

a system of n second-order differential equations, each introducing two constants.

Determination of the constants

1) In the direction of $z > 0$: $\psi_n(z)$ must be regular at infinity (4)

2) In order that the *connecting* conditions for $z = 0$ *keep the same form as before*, we impose the condition

$$[\psi_n(z)]'_{z=0} = 0 \qquad (5)$$

Indeed, $\Sigma \psi_n(z) F_n(r)$ *can then play the same role as* $U(r)$ *for the* B'_m equation which will be modified only by the substitution of $\psi_n(0)$ for α_n, and the A_n equation will keep the same form.

$$\boxed{\text{Perturbation:} \quad \psi(z) = p_0 e^{-p_0 z} \quad \text{in} \quad 0 \leq r \leq R, z > 0}$$

This case corresponds to the rise in heat due to the level of the "plug" of the rise in neutronic flux.

We first consider the general case where p_0 is not equal to a eigenvalue p_n.

The general solution of (3) is of the form:

$$\mu e^{p_n z} + \eta e^{-p_n z}$$

μ and η are determined by the method of "variation of constants".

We thus find:

$$\mu = \mu_0 + \frac{p_0 q_n}{2 p_n (p_0 + p_n)} e^{-(p_0 + p_n)z}$$

$$\eta = \eta_0 - \frac{p_0 q_n}{2 p_n (p_0 - p_n)} e^{-(p_0 - p_n)z}$$

The solution of (3) is hence:

$$\psi_n(z) = \mu_0 e^{+p_n z} + \eta_0 e^{-p_n z} + \frac{p_0 q_n}{2 p_n} \left[\frac{1}{p_0 + p_n} e^{-p_0 z} - \frac{1}{p_0 - p_n} e^{-p_0 z} \right]$$

or

$$\psi_n(z) = \mu_0 e^{p_n z} + \eta_0 e^{-p_n z} + \frac{p_0 q_n}{p_n^2 - p_0^2} e^{-p_0 z}$$

in which μ_0 and η_0 must be determined by the conditions (4) and (5).

Condition (4) implies:

$$\mu_0 = 0$$

Condition (5) implies:

$$\frac{d\psi_n(z)}{dz} = -p_n\eta_0 e^{-p_n z} - \frac{p_0^2 q_n}{p_n^2 - p_0^2} e^{-p_0 z} = 0 \quad \text{for} \quad z = 0$$

that is:

$$\eta_0 = \frac{p_0^2 q_n}{p_n(p_0^2 - p_n^2)}$$

The solution of equation (5) is hence finally:

$$\psi_n(z) = \frac{p_0 q_n}{p_0^2 - p_n^2} \left[\frac{p_0}{p_n} e^{-p_n z} - e^{-p_0 z} \right] \tag{6}$$

Relation between $\varphi(n)$ and $U(r)$

The expansion of $U(r)$ in the $F_n(r)$ has lead to the expression:

$$U(r) = \sum_n \alpha_n F_n(r)$$

where

$$\alpha_n = \int_0^S U(r) G_n(r) r \, dr = \frac{Q}{K} \frac{J_1(p_n R)}{p_n^3 N(p_n)} R$$

The expansion of $\varphi(r)$ in the $F_n(r)$ leads to:

$$\varphi(r) = \sum_n q_n F_n(r)$$

with

$$q_n = \int_0^S \varphi(r) G_n(r) r \, dr = \int_0^R \frac{Q}{K} G_n(r) r \, dr \quad \text{since} \quad \varphi(r) = \begin{cases} 0 \text{ in } (R, S) \\ \dfrac{Q}{K} \text{ in } (0, R) \end{cases}$$

that is:

$$q_n = \frac{Q}{K} \int_0^R \frac{J_0(p_n r)}{N(p_n)} r \, dr = \frac{Q}{K} \frac{J_1(p_n R)}{p_n N(p_n)} R$$

we have hence:

$$q_n = \alpha_n p_n^2 \tag{7}$$

and the solution of (3) is

$$\psi_n(z) = \frac{\alpha_n p_0 p_n}{p_0^2 - p_n^2} [p_0 e^{-p_n z} - p_n e^{-p_0 z}] \tag{8}$$

8*

Transition conditions for z = 0

1)
$$\frac{\partial T}{\partial z} = \sum_n e_n(z) F_n(r) \, z = 0$$

$$\sum_n \left[\frac{\partial \psi_n(z)}{\partial z} + A_n p_n e^{-p_n z} \right] F_n(r) = \sum_n e_n(z) F_n(r) \quad \text{for} \quad z = 0$$

leads to the relation of the same form as in the case without perturbation,

since $\left[\dfrac{\partial \psi_n}{\partial z} \right]_{z=0} = 0$

$$\underline{A_n p_n = e_n(0)} \tag{9}$$

2)
$$\underline{\sum_n u_{nm} B'_m = \psi_n(0)} \tag{10}$$

where:

$$\psi(0) = \frac{\alpha_n}{\dfrac{1}{p_0} + \dfrac{1}{p_n}} \tag{10'}$$

Case where p_0 is equal to an eigenvalue p_n

We then have:

$$\begin{cases} \mu' = -\dfrac{q_n}{2} e^{-2p_n z} \\[3mm] \eta' = \dfrac{q_n}{2} \end{cases} \qquad \begin{cases} \mu = \mu_0 + p_n q_n e^{-2p_n z} \\[3mm] \eta = \eta_0 + \dfrac{q_n}{2} z \end{cases}$$

whence

$$\psi(z) = \mu_0 e^{-p_n z} + p_n q_n e^{-p_n z} + \eta_0 e^{-p_n z} + \frac{q_n}{2} z e^{-p_n z}$$

We also have: $\mu_0 = 0$ and η_0 is given by:

$$\left[\frac{\partial \psi_\mu}{\partial z} \right]_{z=0} = \left[-p_n^2 q_n - p_n \eta_0 + \frac{q_n}{2} \right] e^0 = 0$$

that is:

$$\eta_0 = \frac{q_n}{2 p_n} - p_n q_n$$

We have finally, taking (7) into account

$$\boxed{\psi_n(z) = \frac{\alpha_n p_n}{2} [1 + p_n] e^{-p_n z}} \tag{11}$$

$$p_0 = p_n$$

On the other hand we note that:

$$\psi_n(0) = \frac{\alpha_n p_n}{2}$$

the value assumed by (10') for $p_0 = p_n$

The transition conditions (9), (10), (10') are hence valid in all cases.

Perturbation: $\psi(z)\,\varphi(r) = Q[1 - \lambda I_0(Kr)]$ in $0 \leq r \leq R, z > 0$

---

with

$$\int_0^R [1 - \lambda I_0(Kr)]\, r\, dr = 0 \quad \text{that is} \quad \lambda = \frac{KR}{2I_1(KR)}$$

This case corresponds, with respect to the initial particular case, to the fine description of the heat sources in the multiplying medium (addition of a zero term in the mean).

We thus expand $Q[\ ]$ in the system of the $F_n(r)$

$$Q[1 - \lambda I_0(Kr)] = \sum_n X_n F_n(r) \tag{12}$$

that is:

$$Q\left[1 - \frac{KR}{2}\frac{I_0(Kr)}{I_1(KR)}\right] = \sum_n X_n F_n(r). \tag{12'}$$

We have thus:

$$X_n = \int_0^R Q\left[1 - \frac{KR}{2}\frac{I_0(Kr)}{I_1(KR)}\right] G_n(r)\, r\, dr$$

$$X_n = \frac{Q}{N(p_n)} \int_0^R \left[1 - \frac{KR}{2}\frac{I_0(Kr)}{I_1(KR)}\right] J_0(p_n r)\, r\, dr$$

$$X_n = \frac{Q}{N(p_n)}\left[\frac{R}{p_n} J_1(p_n R) - \frac{KR}{2I_1(KR)}\int_0^R I_0(Kr) J_0(p_n r)\, r\, dr\right]$$

This last integral is a Lommel integral, hence the solution is:

$$\int_0^R I_0(Kr) J_0(p_n r)\, r\, dr = \frac{R}{K^2 - p_n^2}\,[p_n I_0(KR) J_1(p_n R) - K J_0(p_n R) I_1(KR)]$$

Hence we have finally:

$$X_n = \frac{QRJ_1(p_nR)}{p_nN(p_n)}\left[1 + \frac{KRp_n^2}{2(K^2 - p_n^2)}\frac{I_0(KR)}{I_1(KR)} - \frac{K^2Rp_n}{2(K^2 - p_n^2)}\frac{I_0(p_nR)}{I_1(p_nR)}\right]$$

$$\tag{13}$$

We note that the value of X_n is constant.

Equation (3) is hence here:

$$\psi_n''(z) - p_n^2\psi_n(z) + X_n = 0 \quad \text{with} \quad X_n = \text{const} \tag{14}$$

The general solution is:

$$\mu e^{p_nz} + \eta e^{-p_nz}$$

with:

$$\mu = \mu_0 + \frac{X_n}{2p_n^2}e^{-p_nz}$$

$$\eta = \eta_0 + \frac{X_n}{2p_n^2}e^{p_nz}$$

that is:

$$\psi_n(z) = \mu_0 e^{p_nz} + \eta_0 e^{-p_nz} + \frac{X_n}{p_n^2}$$

The conditions (4) and (5) then lead to:

$$\mu_0 - \eta_0 = 0$$

whence

$$\psi_n(z) = \frac{X_n}{p_n^2} = \text{const} \tag{15}$$

The transition conditions for $z = 0$ are hence:

$$A_np_n = e_n(0)$$
$$\sum_m u_{nm}B_m = \frac{X_n}{p_n^2} \tag{16}$$

Generalization of the method

Consider a problem of neutronics (with a group of monokinetic neutrons) characterized by the diagram below, with the following conditions:

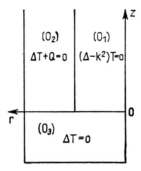

Figure 16

a) regular function for infinite z and $r = 0$.

b) limit conditions:

1) reflexion on the exterior surface

2) continuity of the fluxes and of the neutronic currents on the inter-

faces $D_i \dfrac{\partial Ti}{\partial n}$

The neutronic flux will be:

$$\Phi = \underset{\text{at infinity}}{\theta(r)} + \sum_n A_n e^{-p_n z} F_n(r)$$

And we get, for infinite z:

$$\begin{cases} \theta_1 = \xi I_0(Kr) \\[2mm] \theta_2 = \alpha + \beta \log r - \dfrac{Qr^2}{4} \end{cases}$$

The constants ξ, α, β are determined by the three limit conditions, of reflexion and of continuity of the fluxes and of the currents for $r = S$ and R.

For $z > 0$ and finite:

In the medium (O_1);

$$(\Delta - K^2)\Phi_1 = \left(\dfrac{d^2\theta_1}{dr^2} + \dfrac{1}{r}\dfrac{d\theta_1}{dr} - K^2\theta_1\right)$$

$$+ \sum_n A_n e^{-p_n z}[\Delta_r F_n + p_n^2 F_n - K^2 F_n]$$

In the medium (O_2):

$$\Delta\Phi_2 + Q = \left(\frac{d^2\theta_2}{dr^2} + \frac{1}{r}\frac{d\theta_2}{dr} - Q\right) + \sum_n A_n e^{-p_n z}[\Delta_r F_n + p_n^2 F_n]$$

where θ_1 and θ_2 are the solutions for infinite z, so that the first term in parenthesis in each right-hand side is zero.

F_n is hence a solution of the equations:

$$\begin{cases} \text{in } (O_1): & F_n(1) \to \Delta_r F_n + p_n^2 F_n - K^2 F_n = 0 \\ \text{in } (O_2): & F_n(2) \to \Delta_r F_n + p_n^2 F_n = 0 \end{cases}$$

We set: $p_n^2 - K^2 = l_n^2$ and the required functions are:

in (1): $\lambda_n J_0(l_n r)$

(which is regular for $r = 0$, we have
hence eliminated the solutions with Y_0)

in (2): $\mu_n J_0(p_n r) + \gamma_n Y_0(p_n r)$

The eigenequation will be obtained from the limit conditions for $r = R$ and $r = S$, as above.

There can exist eigenvalues p_n for which l_n is complex, which introduces Bessel functions of the second type, and requires a more extensive study. We note that λ_n, μ_n and ν_n are, here also, completely determined by a normalization condition.

For $z < 0$ (medium O_3) In order to simplify the transition equations, we omit the asymptotic solution $\theta(r)$. We must hence solve:

$$\Delta T = e(r) \begin{cases} = 0 & \text{for } R < r < S \\ = -K^2 \xi I_0(Kr) & \text{for } 0 < r < R \end{cases}$$

We expand $e(r)$ in the system of the $F_m(r)$:
We wish to find:

$$T_3 = \sum_m b_m(z)\, F_m(r)$$

that is:

$$e(r) = \sum_m e_m F_m(r)$$

and:

$$b_m''(z) - q_m^2 b_m(z) = e_m \quad \text{for} \quad m \neq 1$$

with a particular case because the first eigenvalue $q_1 = 0$.

We are lead to expanding $e(r)$
$$\begin{cases} = 0 & \text{in } (R, S) \\ = I_0(Kr) & \text{in } (0, R) \end{cases}$$

In order to expand such a function, we need *a very large number of terms,* which is numerically less satisfying, if we wish to have a good convergence.

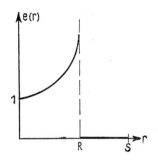

Figure 17

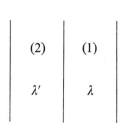

Figure 18

Parabolic problems

The method can be applied to a problem of type:

$$\begin{cases} \lambda \dfrac{\partial T}{\partial t} = \Delta T + \dfrac{Q}{K} & \text{in (1)} \\[2mm] \lambda' \dfrac{\partial T}{\partial t} = \Delta T & \text{in (2)} \end{cases}$$

(λ and λ' are the conductivities of the two media) extended to two infinite concentric cylinders.

We suppose there are limit conditions analogous to those of the preceding problem (but without taking z into account); continuity and regularity for $r = 0$, reflexion for $r = S$, continuity of the fluxes and the currents for $r = R$.

We take the initial solution $T = \theta(r) = T(r, 0)$ for $t = 0$.

We wish to find $T(r, t)$ in the form:

$$T(r, t) = \frac{\sqrt{2}}{S} \varphi_1(t) + \sum_n \varphi_n(t) F_n(r)$$

according to the same principle, we expand:

$$\Lambda \frac{\partial T}{\partial t} - C = \frac{\sqrt{2}}{S} u_1(t) + \sum_n u_n(t) F_n(r)$$

where: $\qquad \Lambda = \begin{cases} \lambda & \text{in (1)} \\ \\ \lambda' & \text{in (2)} \end{cases} \qquad\qquad C = \begin{cases} \dfrac{Q}{K} & \text{in (1)} \\ \\ 0 & \text{in (2)} \end{cases}$

The initial solution implies:

$$\theta(r) = T(r, 0) = \frac{\sqrt{2}}{S}\varphi_1(0) + \sum_n \varphi_n(0)\, F_n(r)$$

We then expand:

$$\theta(r) = \alpha_1 + \sum_n \alpha_n F_n(r) \quad \text{(where } \alpha_1, \alpha_n \text{ are known)}$$

and we set:

$$\varphi_n(t) = A_n + B_n \psi_n(t)$$

where A_n and B_n are determined from the α and the φ, we get

$$\psi'_m(t) = \sum_n u_{nm}\psi_n(t)$$

a linear differential system whose solutions are of the form

$$y_{nj}e^{\frac{t}{x_j}}$$

where the X_j are the eigenvalues of the matrix $\|U_{nm}\|$

We have

$$\sum_m u_{nm} Y_{mj} = X_j Y_{n,j}$$

$$\psi_n(t) = \sum_j k_j Y_{n,\, j} e^{\frac{t}{x_j}}$$

the k_j are determined by the conditions for $t = 0$, that is:

$$\varphi_n(0) = A_n + B_n \psi_n(0) = \alpha_n$$

$$\psi_n(0) = \frac{\alpha_n - A_n}{B_n}$$

and:

$$\sum_j k_j Y_{n,j} = \frac{\alpha_n - A_n}{B_n} \qquad \text{whence the } k_j.$$

Remark 1   It is, in fact, in our interest to use here the Laplace *transform*. In the case where $\lambda = \lambda' = 1$, the differential equations separate for $n = 1, 2, 3$. We prove also that the eigenequation is the same as that which gives the poles in the inverse Laplace transform, and the expansions are indeed the same.

Remark 2   We can solve this problem in the case where we have a finite cylinder. We first perform on "t" the Laplace transformation, which leads to a problem with partial derivatives in two variables r, z. The solutions are expressed in series of Bessel functions. But then, to return to the initial problem by the inverse transformation, we introduce Bessel functions with complex arguments, since, performing a numerical inversion, the arguments of the Bessel functions will be of the form ξ_n where $\xi_n^2 = n^2\pi^2 + \dfrac{s}{\omega_0}$ and where we give the values $i\omega$ to s, the Laplace variable.

Theoretical justification of the spectral method used for the preceding limit problems

1) *Statement of the problem*

We wish to give, without proof, the justification of the preceding method for limit problems of elliptic type, consequently we consider two bounded open sets Ω_1 and Ω_2 of R^n, with "regular" boundaries Γ_1 and Γ_2.

We denote by Δ the Laplace operator

$$\Delta = \frac{\partial^2}{\partial x_1^2} + \cdots + \frac{\partial^2}{\partial x_n^2}$$

and by K_1 and K_2 constants characterizing Ω_1 and Ω_2.

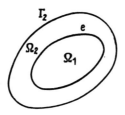

Figure 19

Problem   To find a function U_1 in Ω_1 and U_2 in Ω_2 which satisfies:

$$\begin{cases} -k_1\Delta u_1 = \xi u_1 + f_1 & (1,1) \\ -k_2\Delta u_2 = \eta u_2 + f_2 & (1,2) \end{cases}$$

f_1 and f_2 are given functions in Ω_1 and Ω_2, and ξ and η are real or complex constants which are *not given*.

The limit conditions are as follows:

$$\text{on } \Gamma_1: -k_1 \frac{\partial u_1}{\partial n} = \beta_1(u_1 - u_2) \tag{1,3}$$

$$\text{on } \Gamma_1: k_2 \frac{\partial u_2}{\partial n} = \beta_1(u_2 - u_1) \quad \beta_1 = \text{const} \tag{1,4}$$

where $\dfrac{\partial}{\partial n}$ is taken to be normal exterior to Ω_1.

$$\text{on } \Gamma_2: k_2 \frac{\partial u_2}{\partial n} + \beta_2 u_2 = 0 \quad \beta_2 = \text{const} \tag{1,5}$$

This is a limit problem of an elliptic nature where we search for the spectrum(ξ and η are not given). We will restrict ourselves to the study of the spectrum, that is to say, we will suppose the f_i are zero.

Consider two "regular" functions V_1 and V_2, defined respectively in Ω_1 and Ω_2.

Multiplying (1,1) by $\bar{V}_1$ and integrating in Ω_1 we get:

$$-k_1 \int_{\Omega_1} \Delta u_1 \bar{V}_1 \, dx = \xi \int_{\Omega_1} u_1 \bar{V}_1 \, dx \quad dx = dx_1 \cdots dx_n \tag{1,6}$$

and similarly:

$$-k_2 \int_{\Omega_2} \Delta u_2 \bar{V}_2 \, dx = \eta \int_{\Omega_2} u_2 \bar{V}_2 \, dx \tag{1,6'}$$

If we denote by $d\gamma_1$ and $d\gamma_2$ the surface measures on Γ_1 and Γ_2, Green's formula gives:

$$-k_1 \int_{\Omega_1} \Delta u_1 \bar{V}_1 \, dx = -k_1 \int_{\Gamma_1} \frac{\partial u_1}{\partial n} \bar{V}_1 \, d\gamma_1 + k_1 \int_{\Omega_1} \operatorname{grad} u_1 \operatorname{grad} \bar{V}_1 \, dx$$

and similarly:

$$-k_2 \int_{\Omega_2} \Delta u_2 \bar{V}_2 \, dx = k_2 \int_{\Gamma_1} \frac{\partial u_2}{\partial n} \bar{V}_2 \, d\gamma_1 - k_2 \int_{\Gamma_2} \frac{\partial u_2}{\partial n} \bar{V}_2 \, d\gamma_2$$

$$+ k_2 \int_{\Omega_2} \operatorname{grad} u_2 \operatorname{grad} \bar{V}_2 \, dx$$

(the signs of the integrals on Γ_1 and Γ_2 are determined by the choice of the direction of the normals to the boundaries).

If we set:

$$\pi(u, V) = k_1 \int_{\Omega_1} \operatorname{grad} u_1 \operatorname{grad} \bar{V}_1 \, dx + k_2 \int_{\Omega_2} \operatorname{grad} u_2 \operatorname{grad} \bar{V}_2 \, dx$$

$$- k_1 \int_{\Gamma_1} \frac{\partial u_1}{\partial n} \bar{V}_1 \, d\gamma_1 + k_2 \int_{\Gamma_1} \frac{\partial u_2}{\partial n} \bar{V}_2 \, d\gamma_1 - k_2 \int_{\Gamma_2} \frac{\partial u_2}{\partial n} \bar{V}_2 \, d\gamma_2 \qquad (1,8)$$

which can be written, using the limit conditions (1,3), (1,4), (1,5):

$$\pi(u, V) = k_1 \int_{\Omega_1} \operatorname{grad} u_1 \operatorname{grad} \bar{V}_1 \, dx + k_2 \int_{\Omega_2} \operatorname{grad} u_2 \operatorname{grad} \bar{V}_2 \, dx$$

$$+ \beta_1 \int_{\Gamma_1} (u_1 \bar{V}_1 - u_2 \bar{V}_1 - u_1 \bar{V}_2 + u_2 \bar{V}_2) \, d\gamma_1 + \beta_2 \int_{\Gamma_2} u_2 \bar{V}_2 \, d\gamma_2$$

where $\pi(u, V)$ represents the sum of the left-hand sides of the equation (1,6) and (1,6)', we deduce:

$$\boxed{\pi(u, V) = \xi \int_{\Omega_1} u_1 \bar{V}_1 \, dx + \eta \int_{\Omega_2} u_2 \bar{V}_2 \, dx} \qquad (1,7)$$

In $\pi(u, V)$ u represents the pair $\{u_1, u_2\}$
V represents the pair $\{V_1, V_2\}$

We see thus that if $u = \{u_1, u_2\}$ is a solution of the initial problem, then we have (1,7) for any V_1 and V_2.

Conversely, if $\{u = u_1, u_2\}$ is a pair of regular functions, in the open sets Ω_1 and Ω_2 and satisfying the equation (1,7) for any V_1 and V_2, then u_1 and u_2 are solutions of the problem (1,1), $\cdots$, (1,5).

We are thus reduced to the solution of (1,7) on condition that we give it a precise meaning, thus specifying the function spaces in which we look for the u_i.

2) *The function spaces we use*

We denote by $L^2(\Omega)$ the space of square-summable functions on Ω (an arbitrary open set of R^n).

With the usual scalar product $(u, V) = \int_{\Omega} u(x) \bar{V}(x) \, dx$, we have a Hilbert space.

We then consider the product space $L^2(\Omega_1) \times L^2(\Omega_2)$ with $u_i \in L^2(\Omega_i)$.

If u and V are two elements of this space, we set:

$$\theta(u, V) = \theta_1 \int_{\Omega_1} u_1 \bar{V}_1 \, dx + \theta_2 \int_{\Omega_2} u_2 \bar{V}_2 \, dx \qquad (2,1)$$

where θ_1 and θ_2 are positive constants. We thus endow the space $L^2(\Omega_1)$ $\times L^2(\Omega_2)$ with a Hilbert space structure. We have the usual scalar product only if $\theta_1 = \theta_2 = 1$; we then introduce $H^1(\Omega_1)$, the space of functions $u_1 \in L^2(\Omega_1)$ but which, moreover, have derivatives, in the distributive sense, such that:

$$\frac{\partial u_1}{\partial x_i} \in L^2(\Omega_1) \tag{2,2}$$

Then if u_1 and V_1 are functions in this subspace of $L^2(\Omega)$, we set

$$(u_1, V_1)_{H^1(\Omega)} = \int_{\Omega_1} u_1 \bar{V}_1 \, dx + \int_{\Omega_1} \text{grad } u_1 \text{ grad } \bar{V}_1 \, dx \tag{2,3}$$

We thus endow H^1 with a *Hilbert space structure*.
We introduce similarly the product space $H^1(\Omega_1) \times H^1(\Omega_2)$ in which we define the scalar product by:

$$(u_1, V_1)_{H^1(\Omega_1)} + (u_2, V_2)_{H^1(\Omega_2)}$$

For *regular boundaries* (condition which we will denote (2,4)) (finite unions of manifolds of dimension $(n-1)$, once continuously differentiable, and not tangent at their intersections), for every function of H^1 we can uniquely define the function u_{1/Γ_1}: the value of u_1 on Γ_1; u_{1/Γ_1} is in $L^2(\Gamma_1)$ (square-summable on Γ_1 for the surface measure of $d\gamma_1$).
We have the property:

$$\int_{\Gamma_1} |u_1|^2 \, d\gamma_1 \leq C_1 \|u_1\|^2_{H^1(\Omega_1)}$$

Similarly for u_2 we can uniquely define a function which takes the values u_{2/Γ_2} and u_{2/Γ_2}, elements of $L^2(\Gamma_2)$ and $L^2(\Gamma_1)$ with:

$$\int_{\Gamma_2} |u_2|^2 \, d\gamma_2 \leq C_2 \|u_2\|^2_{H^1(\Omega_2)}$$

It follows, *with the hypothesis (2,4) that $\pi(u, V)$ (cf. (1,8)) is meaningful for u and V in* $H^1(\Omega) \times H^2(\Omega)$ and that moreover

$$\pi(u, V) \leq C_2 \, |||U_2||| \cdot |||V_2|||$$

where the right-hand side represents the norms in the product space $H^1(\Omega_1)$ $\times H^1(\Omega_2)$.

Problem   The spectral problem reduces then to finding numbers p^2, for complex p, such that there exists a non-zero u in $H^1(\Omega_1) \times H^1(\Omega_2)$ satisfying:

$$\pi(u, V) = p^2\theta(u, V)$$ (2,5)

for every element V of $H^1(\Omega_1) \times H^1(\Omega_2)$.

We see that u satisfies all the equations set at the beginning as well as all the limit conditions.

3) Complete systems of eigenfunctions

Fundamental theorem.   *We suppose that the open sets Ω_i are bounded; we suppose that the functions are regular and satisfy the condition (2,4); we suppose that the k_i and β_i are positive.*

Under these conditions, the equation (2,5) admits as its only solution the zero solution, except for a countable set of values of p^2:

$$0 \leq p_1^2 \leq p_2^2 \leq \cdots p_n^2 \leq \cdots$$

$p_n \to \infty$ when $\to \infty$, where each eigenvalue p_n^2 is written as often as its multiplicity.

If we denote by $w^n = \{w_1^n, w_2^n\}$ a solution of (2,5) corresponding to an eigenvalue $p = p_n$ (w is determined up to a multiplicative factor), and if we determine w^n completely by the condition $\theta(w_n, w_n) = 1$, the system of the w^n is *a complete orthonormal system* in the product space $L^2(\Omega_1) \times L^2(\Omega_2)$ (for the scalar product $\theta(u, V)$) and the system $\dfrac{w^n}{p^n}$ is *a complete orthonormal system* in the product space $H^1(\Omega_1) \times H^1(\Omega_2)$ (for the scalar product $\theta(u, V)$).

Practical significance of the freedom of choice of $\theta(u, V)$.

Equation (2,5) implies

$$- k_1 \Delta w_1^{(n)} = \theta_1 p_n^2 w_1^{(n)}$$ (3,1)

$$- k_2 \Delta w_2^{(n)} = \theta_2 p_n^2 w_2^{(n)}$$ (3,2)

on top of the conditions (1,3), (1,4) and (1,5).

In practice there are two choices which are more natural than the others:

1) $\theta_1 = k_1$ and $\theta_2 = k_2$

Under these conditions

$$\begin{cases} -\Delta w_1^{(n)} = p_n^2 w_1^{(n)} & (3,3) \\ -\Delta w_2^{(n)} = p_n^2 w_2^{(n)} & (3,4) \end{cases}$$

We have thus the same eigenvalues, but the scalar product in $L^2(\Omega_1) \times L^2(\Omega_2)$ is:

$$\theta(u, V) = k_1 \int_{\Omega_1} u_1 \bar{V}_1 \, dx + k_2 \int_{\Omega_2} u_2 \bar{V}_2 \, dx$$

which is not the usual scalar product (unless $k_1 = k_2 = 1$). If we wish however to use the usual scalar product, then the system of the w_n is no longer orthogonal, but we can easily associate a biorthogonal system to it.

2) $\theta_1 = 1$ and $\theta_2 = 1$

we introduce thus the usual scalar product on the product space; we then get

$$\begin{cases} -\Delta w_1^{(n)} = \dfrac{1}{k_1} p_n^2 w_1^{(n)} & (3,5) \\[4mm] -\Delta w_2^{(n)} = \dfrac{1}{k_2} p_n^2 w_2^{(n)} & (3,6) \end{cases}$$

and the eigenvalues are distinct (except if $k_1 = k_2$). For f_1 and f_2 not equal to zero (suppose them in $L^2(\Omega_i)$, we wish to find a function u in the product space $H_1(\Omega_1) \times H_1(\Omega_2)$ such that

$$\pi(u, V) = p^2 \theta(u, V) + \int_{\Omega_1} f_1 \bar{V}_1 \, dx + \int_{\Omega_2} f_2 \bar{V}_2 \, dx \qquad (3,7)$$

for all V in $H_1(\Omega_1) \times H_1(\Omega_2)$.

We then have the Riesz-Fredholm alternative

The problem (3,7) admits a unique solution except when p^2 is a eigenvalue. If $p^2 = p_n^2$, a characteristic value, in order to have a solution *we must have:*

$$\int_{\Omega_1} f_1 \bar{w}_{1j} \, dx + \int_{\Omega_2} f_2 \bar{w}_{2j} \, dx = 0$$

for the w_j corresponding to the eigenvalue p_n^2; the solution is then defined modulo these w_j.

Generalization of the theorem

1) We keep the scalar product $\pi(u, V)$ and the open sets Ω_i, but we change the space $H^1(\Omega_1) \times H^1(\Omega_2)$ in order to have *other limit conditions*. This is the case for instance when we go from the heat conditions to the neutronic conditions.

2) We still consider two open sets (Ω_1) and (Ω_2) but in different positions. We keep $\pi(u, V)$. We can consider an arbitrary number of open sets fitting into one another.

3) We modify $\pi(u, V)$ in order to obtain analogous results for second-order *differential systems*. This is the case for diffusion equations with several groups of neutrons.

4) We modify $\pi(u, V)$ in order to obtain systems of arbitrary order *with varying coefficients*. For the higher-order derivatives we are lead to introducing spaces other than $H^1 \cdots$

When the *coefficients* are *varying*, we have the same types of formulas, but in general in this case, only the numerical solution is possible.

4) General problem

We must justify the method when we introduce "above" the bounded open sets an additional dimension (the real variable z).

We then set ourselves in $R_x^n \times R_z$.

Ω_1 and Ω_2 are the bases of the cylinders 0_1 and 0_2, Ω_3 is the basis of 0_3.

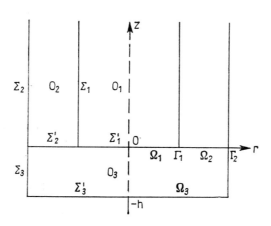

Figure 20

With respect to z, we will have different conditions on both sides of the open sets under consideration. The notations are given by the diagram.

The problem amounts to:

$$- k_i \Delta u_i = f_i \quad \text{in} \quad 0_i$$

where:

$$\Delta = \text{Laplacian in } R_x^n \times R_z$$

and the limit conditions

$$- k_1 \frac{\partial}{\partial n} u_1 = \beta_1(u_1 - u_2) \quad \text{on } (\Sigma_1)$$

$$k_2 \frac{\partial}{\partial n} u_2 = \beta_1(u_2 - u_1) \quad \text{on } (\Sigma_1)$$

$$k_2 \frac{\partial}{\partial n} u_2 + \beta_2 u_2 = 0 \quad \text{on } (\Sigma_2)$$

$$k_1 \frac{\partial}{\partial z} u_1 = \beta_1'(u_1 - u_3) \quad \text{on } (\Sigma_1')$$

$$- k_3 \frac{\partial}{\partial z} u_3 = \beta_1'(u_3 - u_1) \quad \text{on } (\Sigma_1')$$

$$u_2 = u_3 \quad \text{on } (\Sigma_2')$$

$$k_2 \frac{\partial}{\partial z} u_2 - k_3 \frac{\partial}{\partial z} u_3 = 0 \quad \text{on } (\Sigma_2')$$

$$k_3 \frac{\partial}{\partial z} u_3 + \beta_3 u_3 = 0 \quad \text{on } (\Sigma_3)$$

$$\frac{\partial}{\partial z} u_3 = 0 \quad \text{on } (\Sigma_3')$$

We must add conditions for regularity at infinity for u_1 and u_2.

We transform the problem under consideration by integration by parts.

We introduce $H^1(0_i)$ where:

$$u_i \in H^1(0_i) \text{ is equivalent to } \begin{cases} u_i \in L^2(0_i) \\ \dfrac{\partial u_i}{\partial x_j} \in L^2(0_i) \, j = 1, n \\ \dfrac{\partial u_i}{\partial z} \in L^2(0_i) \end{cases}$$

a Hilbert space for the norm with square

$$\int_{0_i} \left\{ |u_i|^2 + \sum_j \left| \frac{\partial u_i}{\partial x_j} \right|^2 + \left| \frac{\partial u_i}{\partial z} \right|^2 \right\} dx \, dz$$

We introduce the space of triplets $u = \{u_1, u_2, u_3\}$ such that $u_i \in H^1(0_i)$. We denote by V the (closed) subspace of

$$H^1(0_1) \times H^1(0_2) \times H^1(0_3)$$

of the functions v such that $V_2 = V_3$ on Σ_2' (guaranteeing the transition condition on this surface).

We wish to find functions u in V.

We will have here for u: solution of the problem and v in V:

$$- k_1 \int\limits_{0_1} \Delta u_1 \bar{V}_1 \, dx \, dz - k_2 \int\limits_{0_2} \Delta u_2 \bar{V}_2 \, dx \, dz - k_3 \int\limits_{0_3} \Delta u_3 \bar{V}_3 \, dx \, dz = (f, V)$$

where

$$(f, V) = \sum_i \int\limits_{0_i} f_i \bar{V}_i \, dx \, dz \quad i = 1, 2, 3.$$

We transform the first term by means of Green's formula, and taking the limit conditions into account, we get:

$$\boxed{a(u, V) = (f, V)} \tag{4.1}$$

where:

$$a(u, V) = \sum_{i=1,2,3} k_i \int\limits_{0_i} \text{grad } u_i \, \text{grad } \bar{V}_i \, dx \, dz$$

$$+ \beta_1 \int\limits_{\Sigma_1} (u_1 \bar{V}_1 - u_2 \bar{V}_1 - u_1 \bar{V}_2 + u_2 \bar{V}_2) \, d\Sigma_1$$

$$+ \beta_2 \int\limits_{\Sigma_2} u_2 \bar{V}_2 \, d\Sigma_2$$

$$+ \beta_1' \int\limits_{\Sigma_1'} (u_1 \bar{V}_1 - u_1 \bar{V}_3 - u_3 \bar{V}_1 + u_3 \bar{V}_3) \, d\Sigma_1'$$

$$+ \beta_3 \int\limits_{\Sigma_3} u_3 \bar{V}_3 \, d\Sigma_3$$

The notations are clear: $d\Sigma_1$ is the area element of Σ_1, hence $d\Sigma_1' = dx$; grad f denotes the vector with components $\dfrac{\partial}{\partial x_j} f, \dfrac{\partial}{\partial z} f$.

Conversely, if a regular u belongs to V, and satisfies (4,1) for all v in V, we verify that u is a solution of the given problem. We are hence led back to the generalized problem:

9*

Problem   We wish to find a u in V which will be a solution of (4,1), for any v in V (V has been defined earlier), we suppose that f_1 is square summable.

We can show that there is a unique solution and we state the:

Theorem (equivalent to the preceding theorem). *We suppose*

$$k_i > 0; \; \beta_i > 0; \; \beta'_i \geqq 0; \; \Omega_1, \Omega_2$$

(*"bases" of the cylinders*) *bounded by regular boundaries* (*condition* (2,4)). *Then the preceding problem admits a unique solution.*

Part two

MONTE CARLO METHODS

CHAPTER 6

The Monte Carlo Method*

The term denoted in 1942, at Los Alamos, a secret file concerning studies by von Neumann.

The Monte Carlo method is a method of solution which consists of applying a numerical process on random numbers.

The method finally reduces to the calculation of "scores" associated with the various events in a game of chance. To obtain a precise result, a large number of plays or "random walks" must be considered. The present development of the method is intimately connected with the progress of the large electronic computers. Of course, we do not toss coins or play dice, but we use series of random numbers. These numbers, which must be uniformly distributed on a given interval (usually [0,1]), can be fabricated by means of physical processes or arithmetical algorithms.

In the first method, we follow a roulette in a casino and we draw up a table of the results. A more sophisticated method would consist of making an electronic ordinator fabricate random numbers, from background noises of electronic nature (e.g. the Mark 1 of Ferranti).

Contrary to what one might think, statistical analysis of the "random" series shows that "chance is not perfect in them". Thus chance was "corrected" in the table of random numbers constructed by the Rand corporation.

In the second method, renouncing the production of truly random numbers, we construct, by means of arithmetical algorithms, pseudo-random sequences which can be verified to satisfy certain statistical tests.

These are not random numbers since the series, which can be very long for certain algorithms, can in fact be predicted and is repetitive, whence the name "pseudo-random numbers". From 1948 till 1954 the "square mean" method was used. For instance, in a sequence of numbers with two figures, each number is given by the median part of the square of the previous number (we remove the first and last figure from the square)

$$(a_1 b_1)^2 = |a_2 b_2| \quad \text{etc} \dots$$

_____

* Some elements of this part are drawn from an I.B.M. Seminar on the Monte Carlo method.

125

The method is not satisfactory since we might find again the initial number or get a double zero before we cover all the two-figure numbers.

The most satisfactory method, and that most commonly used, is that of the modulus, because it allows us, with certain precautions, to construct good series of pseudo-random numbers containing $5 \cdot 10^8$ numbers and more. It defines numbers by means of

$$\varrho_{i+1} = \varrho \varrho_i \ (\mathrm{mod}. \ s^\alpha)$$

where ϱ_0, ϱ and α are given and s is the numeration basis of the computer.

We can find good values of ϱ, ϱ_0 and s.

In the binary system we take

$$s = 2$$

$$\alpha = 37.$$

In fact it is more accurate to speak of "The Monte Carlo methods" since, for a given problem, it will be possible to choose:

a) one numerical process rather than another,

b) the probability distributions of the occuring random numbers.

The Monte Carlo method can be applied to two large categories of problems.

I) Those which involve equations which we plan to solve by constructing by random samplings, an equivalent statistical system. In other words we wish to immerse the given problem into a statistical problem.

II) Those which correspond to the study of a physical system such that it can be described as well as possible by a random sampling (for instance the propagation of a player among the squares of snakes and ladders, the propagation of neutrons, $\cdots$).

Example   Given the functions $A(x)$, $B(x)$ and $k(y, x)$, we propose to calculate the integral:

$$T = \int_0^a A(x) \, U(x) \, \mathrm{d}x$$

where $U(x)$ is a solution of the integral equation:

$$U(x) = \int_0^a U(y) \, k(y, x) \, \mathrm{d}y + B(x)$$

The adjoint system is:

$$\begin{cases} T' = \int_0^a B(y)\, W(y)\, dy \\ \\ W(y) = \int_0^a k(y, x)\, W(x)\, dx + A(y) \end{cases}$$

Since certain transport problems are governed by such systems of integral equations, we will try, conversely, to find a transport model which leads to the same equations as the given system.

Consider a plate of thickness a, which is hit at constant rate by a flux of particles, either at the abscissa of the input face or at the abscissa of the output face.

Let:

$B(x)$ be the amount of particles which have their *first* collision in the interval $(x, x + dx)$.

$k(y, x)\, dx$ be the probability that a particle which had a collision at y will have the next between x and dx.

$U(x)$ be the amount of particles which have their collision between x and $x + dx$.

The difference $U(x) - B(x)$ represents the amount of particles which have a collision in the interval $(x, x + dx)$ after undergoing one or several collisions in the plate. $U(y)\, dy \times k(y, x)\, dx$ of these particles come from the interval $(y, y + dy)$. In total

$$dx \int_0^a k(y, x)\, U(y)\, dy$$

particles have an nth $(n \geq 2)$ collision in the interval $x, x + dx$, whence

$$U(x) = \int_0^a k(y, x)\, U(y)\, dy + B(x)$$

Let $A(x)$ be the probability of a particle coming out of the plate without any collision, from the point x.

Let T be the amount of particles which come out of the plate after at least one collision.

$U(x)$ dx. $A(x)$ of the latter had their last collision in the interval $x, x +$ dx. Hence

$$T = \int_0^a A(x)\, U(x)\, \mathrm{d}x$$

Let $W(y)$ be the probability that a particle which had a collision in y will escape.

This probability is equal to the sum of the probabilities of two independent events, namely:

1) the particle leaves the plate without any new collision. This has probability $A(y)$.

2) the particle leaves the plate after one or more collisions. This has probability:

$$\int_0^a W(x)\, k(y, x)\, \mathrm{d}x$$

So we find the adjoint equation

$$W(y) = \int_0^a k(y, x)\, W(x)\, \mathrm{d}x + A(y)$$

Finally, denoting by T' the amount of particles which leave the plate after at least one collision, we get:

$$T' = \int_0^a B(y)\, W(y)\, \mathrm{d}y$$

Of course $T = T'$; it can be proved analytically by calculating in two different ways the integral

$$\int_0^a U(x)\, W(x)\, \mathrm{d}x$$

Indeed, we get:

$$U(x) = B(x) + K_x U$$
$$W(y) = A(y) + K_y W$$

where K_x and K_y represent respectively the integral operators

$$K_x(U) = \int_0^a U(y)\, k(y, x)\, \mathrm{d}y \quad \text{and} \quad K_y(W) = \int_0^a W(x)\, k(y, x)\, \mathrm{d}x$$

Multiplying the first line by $W(x)$, the second by $U(y)$ and integrating respectively with respect to x and y we get the identity:

$$\int_0^a W(x)\, K_x(U)\, \mathrm{d}x + T' \equiv \int_0^a U(y)\, K_y(W)\, \mathrm{d}y + T$$

Now the two integrals are equal. We deduce:

$$T' = T$$

Having set this, solving the problem by a Monte Carlo method will lead to the following process:

a) We choose a point of first collision, x, by means of the distribution $B(x)$.

b) We then choose at random one of the three following possibilities (by means of random numbers):

1) going out of the plate; probability $A(x)$

2) having a collision between y and $y + \mathrm{d}y$; probability $k(x, y)\,\mathrm{d}y$

3) dying; probability

$$1 - \int_0^a k(x, y)\, \mathrm{d}y - A(x).$$

We continue until the particle disappears or goes out of the plate. We then go on to another particle whose history we will follow in the same manner, etc. $\cdots$

T is evaluated by taking the ratio of the number n of particles which leaves the plate with the number N of all those of which we considered the history.

Statistical error

The preceding statistical process is equivalent to a game of chance in which a random variable t takes the value 1 if the particle leaves the plate, and the value 0 in all other cases. Denoting by $\langle t \rangle$ the average value of t, we get indeed:

$$\langle t \rangle = 1 \mathrm{x} T + 0 \mathrm{x} (1 - T)$$

$$\langle t \rangle = T$$

We also get

$$\langle t^2 \rangle = T$$

since $t^2 = t$.

The mean value, σ^2 of the squared difference

$$(t - T)^2$$

in called the variance of t.

$$\sigma^2 = \langle (t - T)^2 \rangle = \langle t^2 \rangle - 2T\langle t \rangle + T^2$$

$$\sigma^2 = \langle t^2 \rangle - T^2$$

$$\sigma^2 = T(1 - T)$$

This being so, in a first series of random selections we made N choices $t_i (i = 1, 2, \cdots, N)$ of the variable t. We deduce a first statistical value of T, say $\tilde{T}$

$$\tilde{T} = \frac{1}{N} \sum_{i=1}^{N} t_i$$

Repeating a second time, then a third, etc., $\cdots$ the same game of N selections we obtain a series of approximate values of T, say

$$\tilde{T}_1, \tilde{T}_2, \tilde{T}_3, \cdots$$

The error is obviously all the more small as N is large. In order to evaluate the error, the variance $\tilde{\sigma}^2$ of the random variable $\tilde{T}$

$$\tilde{\sigma}^2 = \langle (\tilde{T} - T)^2 \rangle = \langle \tilde{T}^2 \rangle - 2T\langle \tilde{T} \rangle + T^2$$

$$\tilde{\sigma}^2 = \langle \tilde{T}^2 \rangle - T^2$$

will evaluate the moment of second order, noting that the selections t_i are independent:

$$\langle \tilde{T}^2 \rangle = \frac{1}{N^2} \langle \sum_{i=1}^{N} t_i^2 + \sum_{i=1}^{N} t_i \sum_{j=1}^{N} t_j \rangle = \frac{1}{N^2} \{ N\langle t^2 \rangle + NT(N - 1) T \}$$

$$= \frac{1}{N} \langle t^2 \rangle + \frac{N - 1}{N} T^2$$

whence, substituting in the expression for $\tilde{\sigma}^2$

$$\tilde{\sigma}^2 = \frac{1}{N} (\langle t^2 \rangle - T^2) = \frac{\sigma^2}{N}$$

In short, the random variables t and $\tilde{T}$ have the same mean value T but the variance of the second is N times smaller than that of the first.

In particular, the mean relative error in the last case is:

$$\frac{\tilde{\sigma}}{T} = \frac{\sigma}{T\sqrt{N}} = \sqrt{\frac{1-T}{NT}}$$

In many problems of interest (protections, screening, etc $\cdots$) the transmission T (exit probability) is small so that

$$\frac{\tilde{\sigma}}{T} \approx \frac{1}{\sqrt{NT}}$$

In order to evaluate T with a mean relative error $E = 10\%$, we would have to follow

$$N = \frac{1}{E^2} \cdot \frac{1}{T} = \frac{100}{T}$$

particle histories per game; and this number is beyond the possibility of the fastest machines for the values of T under consideration. For instance at

$$10^{-6} < T < 10^{-10} \rightarrow 10^8 < N < 10^{12}$$

In order to counter this major difficulty, we might have to abandon the direct game t, whose disadvantage resides in the fact that we only devote an infinitesimal part of our effort to the particles which are actually transmitted. To define a replacement game, we will be lead to choosing various numbers t^* in a population t^* with a probability law $p(t^*)$ such that

$$\langle t^* \rangle = \langle t \rangle = T$$

but

$$\sigma^{*2} \ll \sigma^2$$

Proceeding in this way we decrease the error for the statistical mean $\tilde{T}^*$

$$\tilde{T}^* = \frac{1}{N} \sum_{i=1}^{N} t_i^*$$

since

$$\left[\frac{\tilde{\sigma}^*}{T} = \frac{\sigma^*}{T\sqrt{N}} \right] \ll \left[\frac{\tilde{\sigma}}{T} = \frac{\sigma}{T\sqrt{N}} \right]$$

Conversely, for a given relative error E, we can decrease the number of histories since:

$$N^* = \left[\frac{1}{E} \cdot \frac{\sigma^*}{T} \right]^2 \ll N = \left[\frac{1}{E} \cdot \frac{\sigma}{T} \right]^2$$

So that we use the *weighting method* which, under certain conditions, allows us theoretically to define a game with zero variance.

The weighting method

We show by examples how this method allows us to:

a) avoid tedious calculations,

b) minimize the variance of the final result.

Example 1   We wish to calculate the integral

$$J = \int_0^1 g(x)f(x)\,\mathrm{d}x$$

where $f(x)$ is a function which is not < 0, which we can always consider as being normalized.

$$\int_0^1 f(x)\,\mathrm{d}x = 1$$

and where $g(x)$ is a function ≥ 0.

In probabilistic language, J is the mean value on $(0,1)$ of the function $g(x)$ of the random variable x with probability density $f(x)$.

Associating with this statistical scheme a game of chance with same probability, we can set up a correspondence between a series of selections x_i $(i = 1, 2, \cdots, N)$ and a statistical value $\tilde{J}$ of the random function $g(x)$

$$\tilde{J} = \frac{1}{N} \sum_{i=1}^{N} g(x_i)$$

with variance σ^2

$$\sigma^2 = \int_0^1 g^2(x)f(x)\,\mathrm{d}x - J^2$$

If $f(x)$ is a "regular" function (with slow variation), the points x_i will be roughly equidistributed in the interval $(0,1)$ which will thus lead to a poor evaluation of the integral, when the function $g(x)$ presents jumps or a rapid variation. We will avoid a poor use of the Monte Carlo method by choosing the points x_i not "in direct game" but with a probability law *such that the various values of $g(x)$ are assumed with a frequency proportional to the magnitude of these values in the result.*

Consider for instance (cf. figure 21) a function $g(x)$ with a big jump in the neighbourhood of zero

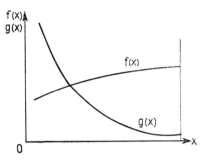

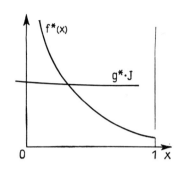

Figure 21   Figure 22

Set

$$J^* = \int_0^1 g^*(x) f^*(x)\, dx$$

with

$$\int_0^1 f^*(x)\, dx = 1$$

and

$$g^*(x) \equiv \frac{g(x) f(x)}{f^*(x)}$$

so that J and J^* are indeed equal.

Defining a game of chance with probability $f^*(x)$ we can make a statistical value $\tilde{J}^*$ of the random function $g^*(x)$ correspond to a series of N selections x_i $(i = 1, 2, \cdots, N)$.

$$\tilde{J}^* = \frac{1}{N} \sum_{i=1}^{N} g(x_i) \frac{f(x_i)}{f^*(x_i)}$$

with variance $(\sigma^*)^2$

$$(\sigma^*)^2 = \int_0^1 \left(\frac{gf}{f^*}\right)^2 f^*\, dx - J^2$$

We determine $f^*(x)$ so as to minimize the variance. Since J^2 is a constant, this amounts to choosing a function $f^*(x)$ which minimizes the integral

$$\int_0^1 \frac{g^2 f^2}{f^*}\, dx$$

and satisfies the condition

$$\int_0^1 f^* \, dx = 1$$

Introducing a multiplier λ, we can express the preceding conditions by annihilating the variation of the function

$$H[f^*(x), x] \equiv \frac{g^2(x) f^2(x)}{f^*(x)} + \lambda f^*(x)$$

for an arbitrary increase δf^*, and this for any x. Whence

$$\frac{\partial H}{\partial f^*} \equiv - \frac{g^2 f^2}{f^{*2}} + \lambda = 0$$

and

$$f^*(x) = \frac{1}{\sqrt{\lambda}} \cdot g(x) f(x)$$

We determine λ by means of the normalization condition:

$$1 = \int_0^1 f^*(x) \, dx = \frac{1}{\sqrt{\lambda}} \int_0^1 g(x) f(x) \, dx = \frac{J}{\sqrt{\lambda}}$$

We get finally:

$$f^*(x) = \frac{g(x) f(x)}{J}$$

$$g^*(x) = \quad J$$

so that

$$(\sigma^*)^2 = \quad 0$$

Optimally we would hence get figure 22.

The variance of the optimal game is zero and hence the statistical value $\tilde{J}$ coincides with the exact value J. This result is of course purely theoretical since in order to play this game we would have to know J which is precisely the unknown of the problem.

In practice, we can get a good approximation of J by various preliminary methods, or proceed by successive approximations. We begin with a rough value of J, say J_1. We iterate with the probability density $\dfrac{gf}{\tilde{J}_2}$ where $\tilde{J}_2$ is the result of the preceding calculation, etc $\cdots$

Example 2   We return to the calculation of the definite integral

$$T = \int_0^a B(x)\, W(x)\, dx$$

where $W(x)$ is a solution of the integral equation:

$$W(x) = A(x) + \int_0^a k(x, y)\, W(y)\, dy$$

We have seen that when T is very small the direct transport game can be too much for even the fastest machines to cope with, even for a relatively small error (for instance $\Delta T/T = 10\%$).

Maintaining the general statistical interpretation, we are then lead to replacing $B(x)$ by a function $B^*(x)$ representing the amount of particles which, in a modified game, would have their first collision in the interval $x, x + dx$.

$B^*(x)$ as well as $B(x)$ is normalized in $(0, a)$

$$\int_0^a B(x)\, dx = \int_0^a B^*(x)\, dx = 1$$

It is always possible to satisfy this last condition by setting:

$$B^*(x) \equiv \frac{B(x)\, M(x)}{m}$$

where $M(x)$ is an arbitrary function and m is a constant defined by:

$$m = \int_0^a B(x)\, M(x)\, dx$$

With this probability law, the mean value of the function $W^*(x)$ in $(0, a)$ will remain equal to T if

$$W^*(x) \equiv W(x)\, \frac{m}{M(x)}$$

since

$$T = \int_0^a B(x)\, W(x)\, dx = \int_0^a B^*(x)\, W^*(x)\, dx$$

$W^*(x)$ satisfies an equation which can be readily deduced from the first by substitution

$$W^*(x)\frac{M(x)}{m} = A(x) + \int_0^a k(x, y)\frac{M(y)}{m} W^*(y)\,dy$$

Let again, multiplying the two members by $\dfrac{m}{M(x)}$ and introducing the notations

$$\begin{cases} A^*(x) = \dfrac{m\,A(x)}{M(x)} \\[3mm] k^*(x, y) = \dfrac{k(x, y)\,M(y)}{M(x)} \end{cases}$$

$$W^*(x) = A^*(x) + \int_0^a k^*(x, y)\,W^*(y)\,dy$$

We will determine the best value of $M(x)$ by minimizing the variance σ^{*2} of the random score $W^*(x)$ in the probability game $B^*(x)$.

Proceeding as for the first example, we get:

$$\begin{cases} M(x) \equiv W(x) \\[2mm] m \equiv T \end{cases}$$

The optimum game is hence defined by the probabilities

$$B^*(x) = \frac{B(x)\,W(x)}{T}$$

$$A^*(x) = \frac{TA(x)}{W(x)}$$

$$k^*(x, y) = \frac{k(x, y)\,W(y)}{W(x)}$$

We verify that $W^*(x)$ is then a constant equal to T. The variance

$$\sigma^{*2} \equiv \int_0^a W^*(x)^2\,B(x)\,dx - T^2$$

is thus zero and hence the value determined by this Monte Carlo method is equal to the exact value.

Once more this solution is of course completely theoretical since it sup-
poses that the solution $W(x)$ of the integral equation is known, and this is
precisely one of the unknowns of the problem.

In practice we can proceed by iteration starting with an approximate
value of $W(x)$, say $M(x)$. A choice of $M(x)$, no matter how rough, will
always be preferable to the unit echelon, $M(x) \equiv 1$, whose main disadvantage is that it ignores the *magnitude of the contribution of $W(x)$* in the
calculation of the integral T.

Say we wish to calculate the integral

$$T = \int_0^a B^*(x) \frac{m}{M(x)} W(x) \, dx$$

knowing that

$$W(x) = A(x) + \int_0^a k(x, y) \, W(y) \, dy$$

We choose a point x_1 of first collision by means of the probability law
$B^*(x)$. From the preceding, the score connected with this selection will be:

$$t \approxeq \frac{m}{M(x_1)} W(x_1)$$

which is an approximate value of the integral all the more accurate as $B^*(x)$
is well chosen. In particular, for $M(x) \equiv W(x)$, we would have the exact
value T.

We calculate $W(x_1)$ by determining, by a random process, the history of
the particle from x_1 to its exit from the plate.

Let $n(x)$ be the function defined by

$$n(x) \equiv A(x) + \int_0^a k(x, y) \, M(y) \, dy$$

Setting:

$$A^*(x) = \frac{A(x)}{n(x)}$$

$$k^*(x, y) = \frac{k(x, y) \, M(y)}{n(x)}$$

We see that it is possible to define a random process (P) which contains
only two types of events with complementary probabilities, namely:

10*

Case $(\alpha)\ k*(x, y)\ \mathrm{d}y$ = Prob. that a particle which had a collision at x will have the next between y and $y + \mathrm{d}y$.

Case $(\beta)\ A*(x)$ = Prob. that a particle which had a collision at x will go out of the plate without any new collisions.

These events have complementary probabilities, because the probability of death at x is zero:

$$1 - A*(x) - \int_0^a k*(x, y)\ \mathrm{d}y = 0$$

Substituting $A*(x)$ and $k*(x, y)$ in the integral equation satisfied by $W(x)$, we get:

$$W(x) = A*(x) \cdot n(x) + \int_0^a k*(x, y) \frac{n(x)}{M(y)} W(y)\ \mathrm{d}y$$

This equation can be interpreted as giving the *mean value* of $W(x)$ in the random process P on condition that we define the scores in the following manner

$$W(x) \equiv n(x) \qquad \text{(in case } \beta)$$

$$W(x) \equiv \frac{n(x)}{M(y)} W(y) \qquad \text{(in case } \alpha)$$

We will choose between (α) and (β) by comparing a random number ξ with the probabilities of each of the two possible events.

If (β) occurs, the particle will leave the plate directly and we take

$$W(x_1) = n(x_1)$$

whence

$$T \eqsim \frac{m}{M(x_1)} n(x_1)$$

The history of the particle will end at x_1.

If (α) occurs, the particle will undergo a collision which will lead it from x_1 to x_2 and we take

$$W(x_1) = \frac{n(x_1)}{M(x_2)} W(x_2)$$

whence

$$T \eqsim \frac{m}{M(x)} \cdot \frac{n(x_1)}{M(x_2)} W(x_2)$$

We follow the history of the particle deciding, by a new selection, on the nature of the collision at x_2, etc ...

Finally if the particle has gone through the points $x_1, x_2, \cdots, x_n$ before escaping, we determine the random value of the function $W(x_1)$ by means of the system:

$$W(x_1) = \frac{n(x_1)}{M(x_2)} \cdot W(x_2)$$

$$W(x_2) = \frac{n(x_2)}{M(x_3)} \cdot W(x_3)$$

$$\cdots \cdots \cdots \cdots \cdots$$

$$W(x_{n-1}) = \frac{n(x_{n-1})}{M(x_n)} \cdot W(x_n)$$

$$W(x_n) = n(x_n)$$

that is, multiplying side by side:

$$W(x_1) = \frac{n(x_1)}{M(x_2)} \cdot \frac{n(x_2)}{M(x_3)} \cdots \frac{n(x_{n-1})}{M(x_n)} \cdot n(x_n)$$

To $W(x_1)$ there will correspond a first sampling of the integral T, namely:

$$t = \frac{m}{M(x_1)} \cdot \frac{n(x_1)}{M(x_2)} \cdot \frac{n(x_2)}{M(x_3)} \cdots \frac{n(x_{n-1})}{M(x_n)} \cdot n(x_n)$$

Repeating the game a second, then a third time, etc ... we would get a series of approximate values of T namely t_1, t_2, t_3, etc... We will define the statistical value T corresponding to N particle histories as the arithmetical mean of the scores:

$$\tilde{T} = \frac{1}{N} \sum_i t_i$$

We are guaranteed, by the choice of the probability laws $B^*(x)$, $A^*(x)$ and $k^*(x, y)$, a convergence which is all the better as $M(x)$ is closer to $W(x)$. In particular, for $M(x) \equiv W(x)$, the scores are independent of the history under consideration since

$$n(x) \equiv W(x)$$

$$m \equiv T$$

$$t_i \equiv T$$

So we find the optimum game, with zero variance ($\sigma^2 = 0$) in which the statistical value T coincides with the exact value T.

Remark   The existence of methods allowing us to accelerate the convergence of a Monte Carlo computation is connected with the following fact:

Let Φ be a physical transport problem and E the system of equations which characterize it. The relation $\Phi \to E$ is univocal.

On the contrary, several random schemes Φ_i can be associated with E (and among them, of course, is the real, initial problem Φ) which differ from one another only by the mechanics of the propagation, but are equivalent, macroscopically, with respect to certain average values.

The weighting method is based on the indetermination of the multi-vocal relation $X(E) \to X(\Phi_i)$ in which X represents certain unknowns connected with E; it allows us to replace the real scheme Φ, by a scheme Φ' for which there is more rapid convergence to the unknowns X.

In general, the decrease in variance requires a slight increase of numerical computations per game, but, as we have seen, the balance shows a large profit.

CHAPTER 7

Study of the Propagation of Fast Neutrons in Water

Consider a semi-infinite water lamina and a plane source of neutrons with energies between 20 Mev and 300 kw (fission spectrum).

The neutrons are slowed down by elastic scattering with the hydrogen, by elastic or inelastic scattering with the oxygen, before escaping (10 meters) or being slowed down to an energy lower than a given limit value (1 ev).

We propose to determine, among other phenomena, the *slowing-down flux*, that is:

1) The deformation of the fission spectrum at various distances from the entrance face, for each of the initial energy bands and various final energy bands.

2) The albedo for fast neutrons in terms of the direction and the energy of the incident neutrons.

The method consists of following, *collision after collision*, the propagation of a large number of neutrons in the water. But it is possible, by means of a suitable weighting, to associate a neutron with each collision, and to use, in order to deal with certain problems, the cloud of neutrons defined in this way. (The weight attributed to each collision is a strongly decreasing function, as we will see, of the effected trajectory.)

The behaviour of a neutron depends primarily on the nature of the collision.

Below 1 ev the neutron can vanish by (n, γ) reaction with a H nucleus [the oxygen capture is entirely negligible in the thermal range]. To simplify matters we will say that the life of a neutron has ended when the energy falls below $E_e = 1$ ev.

At high energies, the inelastic cross-section of the oxygen is considered as an absorption cross-section.

All the other collisions are elastic, i.e. they preserve both the momentum and the kinetic energy of the pair (n, H) or (n, O).

141

Two elastic collisions can differ, however, by the scattering law in the center of inertia system.

a) in the case of hydrogen, the scattering is isotropic and uniquely forward (because we suppose that the collision is with a nucleus of same the mass as the neutron).

b) in the case of oxygen, the scattering is not isotropic in the neighbourhood of the resonance peaks corresponding to the energies:

$$0,347 \text{ Mev} < E < 0,544 \text{ Mev}$$

and

$$0,897 \text{ Mev} < E < \sim 1 \text{ Mev}$$

At high energies it can be determined from tables; at low energies it is isotropic, but with scattering in all directions (the oxygen being considered as infinitely heavy).

The first collision is of great importance; a fast neutron can loose, indeed, all or part of its initial energy during a single collision with the H. Moreover the mean range decreases very rapidly afterwards, since the scattering cross-section of the H, small at 20 Mev, increases rapidly when the energy decreases. A poor evaluation of the target distribution during the first collision might thus lead to important errors in the study of the propagation of the neutrons. So that the first collision requires special treatement. We first study the ith collision.

CONSTRUCTION OF THE NEUTRON LIVES

Dividing if necessary the source spectrum into constant-energy bands, we can always come back to the case of a source of monokinetic neutrons, of energy E_0.

Distance covered: random variable $\varXi_1$

Let

$M_{i-1}(x_{i-1}, y_{i-1}, z_{i-1})$ be the position of the $(i-1)$th collision

$\vec{\omega}_{i-1}(\alpha_{i-1}, \beta_{i-1}, \gamma_{i-1})$ the direction of path of the neutron after the $(i-1)$th collision

E_{i-1} the energy after the $(i-1)$th collision

ε_T^{i-1} total effective cross-section of the water corresponding to this energy

$r = \left| \overrightarrow{M_{i-1}M} \right|$ the distance covered since the collision $(i-1)$

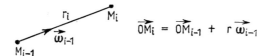

$$\overrightarrow{OM_i} = \overrightarrow{OM_{i-1}} + r\,\vec{w}_{i-1}$$

Figure 23

By the definition of the effective cross-section, the probability $p(r)\,dr$ that the following collision will take place in the interval $r, r + dr$ is

$$p(r)\,dr = \varepsilon_T^{i-1}\,dr\,e^{-\varepsilon_T^{i-1}r}$$

where

$$\int_0^\infty p(r)\,dr = 1$$

To each value of $r \in (0, +\infty)$ we associate the number $\varXi_1 \in (0, 1)$ defined by

$$\varXi_1 = e^{-\varepsilon_T^{i-1}R}$$

Since the function $\varXi_1(R)$ is uniformly decreasing, the correspondence $R, \varXi_1$ is biunivocal. It follows that

$$Pr[\varXi_1 \in (\xi_1, \xi_1 + d\xi_1)] = Pr[R \in (r, r + dr)]$$

where

$$\begin{cases} \xi_1 = e^{-\varepsilon_T^{i-1}r} \\ \\ d\xi_1 = -\varepsilon_T^{i-1}\,e^{-\varepsilon_T^{i-1}r}\,dr \end{cases}$$

We deduce, using that $d\xi_1 > 0 \to dr < 0$

$$Pr[\varXi_1 \in (\xi_1, \xi_1 + d\xi_1)] = +d\xi_1$$

Consequently, choosing a value for $r \in (0, +\infty)$ with the probability distribution $p(r)$ amounts to choosing a value for $\xi_1 \in (0, 1)$ with the uniform probability distribution

$$M(\xi_1) = 1$$

So we can calculate the distance r_i covered between the $(i - 1)$th and the ith collision by means of the relation:

$$r_i = -\frac{1}{\varepsilon_T^{i-1}}\log \xi_1$$

where ξ_1 is a random number equidistributed in the interval $(0, 1)$.

Remark concerning the first collision We require that the first collision occur on the interface $x = 0$, under normal incidence

$$\vec{\omega}_0 \equiv \vec{x} \quad \text{say} \quad \begin{cases} \alpha_0 = 1 \\ \beta_0 = \gamma_0 = 0 \end{cases}$$

$$x_1 = 0 \quad \text{say} \quad [\xi_1] = 1$$

non random   (whence $r_1 = 0$)

The problem under consideration is thus a study of paths parallel up to a translation to the x-axis, of amplitude equal to the first free path of each neutron in the water lamina, namely

$$r_1 = -\frac{1}{\varepsilon_T^0} \log \xi_1$$

where ξ_1 is a random number.

Proceeding thus, we reduce, in the mean, the total distance X covered by the neutrons to be a length equal to

$$\langle r_1 \rangle = -\frac{1}{\varepsilon_T^0} \int_0^1 \log \xi \, d\xi = \frac{1}{\varepsilon_T^0}$$

Since this length is in general rather large (the total cross-section of the water is small at high energies) the treatment under consideration allows us, finally, to give a good representation of the propagation of neutrons in a lamina of finite thickness, x_e, for a given number, N, of neutrons.

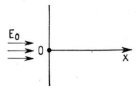

Figure 24

In the problem we have taken $x_e = 10$ m, where the life of a neutron is considered finished when $x > x_e$.

Nature of the collision

Since the medium is homogeneous, the effective cross-section of each of the constituents only depends on the energy. We can hence associate well-determined reaction probabilities with each path segment.

On the path $\overrightarrow{M_{i-1}M_i}$, $E = E_{i-1}$, we will denote by
σ_0 the total effective cross-section of the O
σ'_0 the elastic effective cross-section of the O
σ_H the total effective cross-section of the H (it is in fact the elastic cross-section since $E_{i-1} > 1$ ev).
We set

$$\alpha \equiv \frac{\sigma_0}{\sigma_0 + 2\sigma_H}$$

$$\beta \equiv \frac{\sigma'_0}{\sigma_0 + 2\sigma_H}$$

We get

$$1 - \alpha \equiv \frac{2\sigma_H}{\sigma_0 + 2\sigma_H}$$

The numbers α and $1 - \alpha$ represent respectively the collision probabilities with the O and with the H for the energy E_{i-1}.

Among the collisions with the O we must distinguish the elastic collisions, whose probability in the total balance is β, and the inelastic collisions (we consider them as absorptions) with probability $\alpha - \beta$.

We will hence determine the nature of the collision from a random selection of random numbers ξ_2 equidistributed in the interval $[0, 1]$.

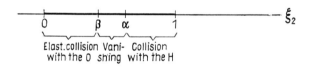

Figure 25

If

$\alpha < \xi_2$ collision with the H

$\beta < \xi_2 < \alpha$ vanishing

$\xi_2 < \beta$ elastic collision with the O

Of course at energies where there is no inelastic collision with the O, we have $\beta = \alpha$.

Study of the first collision

At the first collision the N neutrons of the initial puff have the same energy E_0. So that they all have the same probability $1 - \alpha(E_0)$ of collision with a nucleus of H. There is hence no great advantage in introducing statistically the number N_H of these neutrons since we know, a priori, the exact mean value.

$$N_H = N \cdot [1 - \alpha(E_0)]^*$$

In the same way we can evaluate the number N_O of neutrons whose first collision is with an oxygen nucleus and is elastic

$$N_O = N \cdot \beta(E_0)^*$$

The difference $N - (N_H + N_O)$ is the number of neutrons which vanish at the first collision.

Having established this, numbering from 1 to N the neutrons whose source is in a unique energy band, and denoting by k (modulo N) the number of one of them, we determine the nature of the first collision in the following manner:

$$k \leq N_H \qquad \text{collision with the H}$$

$$N_H < k \leq N_O + N_H \qquad \text{elastic collision with the O}$$

$$N_O + N_H < k \qquad \text{vanishing}$$

η being an random number, we take (cf. the following)

$$\xi_3 = \frac{k - 1 + \eta}{N_H} \qquad \text{for a collision with H}$$

$$\xi_3 = \frac{k - 1 - N_H + \eta}{N_O} \qquad \text{for a collision with O}$$

$$\xi_4 \text{ for arbitrary collisions}$$

Direction and energy after collision: random variables $\varXi_3$ and $\varXi_4$

With the preceding hypotheses, the collisions which are not followed by absorption are elastic for any energy. Moreover, the scattering is isotropic in the center of inertia system, for any energy in the case of hydrogen, and apart from the resonances for the oxygen.

The energy after collision depends on the angle of deflection θ and on the ratio A of the atomic mass of the target nucleus to that of the neutron:

$$\cos \theta = \frac{1}{2}(A + 1)\sqrt{\frac{E_1}{E_0}} - \frac{1}{2}(A - 1)\sqrt{\frac{E_0}{E_1}}$$

---

* We take the integral part.

θ itself depends on the scattering angle Θ in a frame of reference connected with the center of inertia:

$$\cos \theta = \frac{1 + A \cos \Theta}{\sqrt{1 + 2A \cos \Theta + A^2}}$$

Finally, we get directly

$$\frac{E_1}{E_0} = \frac{1 + 2A \cos \Theta + A^2}{(1 + A)^2}$$

These formulas follow from the theorems on conservation of the motion and of the kinetic energy.

Let $\vec{u}$ and $\vec{U}$ be the respective velocities of the neutron and the target nucleus in the center of inertia system. These velocities are colinear since the center of inertia is at rest:

$$0 = \vec{u} + A\vec{U}$$

Hence, introducing the unitary vector $\vec{\Omega}$ of the direction of propagation of the neutron and the number $U > 0$,

$$\vec{u} = AU\vec{\Omega}$$

$$\vec{U} = -U\vec{\Omega}$$

$$|\vec{\Omega}| = 1$$

With these notations, the kinetic energy in the motion about the center of gravity has value, up to a constant factor:

$$\vec{u}^2 + A \cdot \vec{U}^2 = A(A + 1) U^2$$

It only depends on U. Thus there is conservation if $|\vec{U}|$, and hence $|\vec{u}|$, are not altered by the collision. In short, only the direction of the velocities changes.

Denoting respectively by 0 and 1 the magnitudes before and after the collision, we get:

$$\begin{cases} \vec{u}_0 = AU\vec{\Omega}_0 \\ \vec{U}_0 = -U\vec{\Omega}_0 \\ \vec{\Omega}_0^2 = 1 \end{cases} \text{ and } \begin{cases} \vec{u}_1 = AU\vec{\Omega}_1 \\ \vec{U}_1 = -U\vec{\Omega}_1 \\ \vec{\Omega}_1^2 = 1 \end{cases}$$

The variable $\Theta^* = (\vec{\Omega}_0, \vec{\Omega}_1)$ is the angle of deflection in the center of inertia system. When *there is isotropy in all directions*, which is not the case for hydrogen, all angles are equi-probable on the sphere of radius 1. In particular:

$$Pr[\Theta^* \in (\Theta, \Theta + d\Theta)] = \frac{\text{area of the annulus } (\Theta, \Theta + d\Theta)}{\text{area of the sphere}} = \frac{2M \sin \Theta \, d\Theta}{4M}$$

that is:

$$Pr[\Theta^* \in (\Theta, \Theta + d\Theta)] = p(\Theta) \, d\Theta = \tfrac{1}{2} \sin \Theta \, d\Theta$$

To the random variable $\Theta^* \in (0, \pi)$ we associate the random variable $H = \cos \Theta^* [H \in (-1, 1)]$. Since the function $H(\Theta^*)$ is uniformly decreasing, the correspondence Θ^*, H is biunivocal. It follows that:

$$Pr[H \in (\eta, \eta + d\eta)] = Pr[\Theta^* \in (\Theta, \Theta + d\Theta)]$$

where

$$\begin{cases} \eta = \cos \Theta \\ d\eta = -\sin \Theta \, d\Theta \end{cases}$$

We deduce, noting that $d\eta > 0 \rightarrow d\Theta < 0$,

$$Pr[H \in (\eta, \eta + d\eta)] = \tfrac{1}{2} \, d\eta$$

The variable $H = \cos \Theta^*$ is hence equidistributed in the interval $(-1, +1)$. It is convenient to substitute for H the variable $\Xi_3 = \dfrac{1 + H}{2}$ which varies in $(0, 1)$. We readily obtain the probability law of the latter by noting that:

$$Pr[\Xi_3 \in (\xi_3, \xi_3 + d\xi_3)] = Pr[H \in (\eta, \eta + d\eta)]$$

where

$$\begin{cases} \xi_3 = +\tfrac{1}{2}(1 + \eta) \\ d\xi_3 = \tfrac{1}{2} \cdot d\eta \end{cases}$$

We deduce:

$$Pr[\Xi_3 \in (\xi_3, \xi_3 + d\xi_3)] = d\xi_3$$

Consequently, choosing a value of $\Theta \in (0, M)$ with probability distribution $p(\Theta) = \tfrac{1}{2} \sin \Theta$ amounts to choosing a value of $\xi_3 \in (0, 1)$ with the uniform probability distribution $\pi(\xi_3) = 1$.

We will hence calculate the angle of deflection Θ in the center of inertia system, when there is isotropy, by means of the relation

$$\cos \Theta = 2\xi_3 - 1$$

where ξ_3 is a random number equidistributed in the interval $(0, 1)$.

In the case where is anisotropy, the distribution of the random variable $H = \cos \Theta^*$ depends both on the angle and on the energy:

$$Pr[H \in (\eta, \eta + d\eta)] = p(\eta, E)\, d\eta$$

$$\int_{-1}^{+1} p(\eta, E)\, d\eta = 1$$

The values of $p(\eta, E)$ are given in tables with double entries, for the principal elements. They differ all the more from the value $\frac{1}{2}$ as the anisotropy is more pronounced.

This being so, in order to choose η, we will make use not of the probability density $p(\eta, E)$ but of the distribution function $P(\eta, E)$ which is its integral:

$$P(\eta_0, E) \equiv Pr[H < \eta_0] = \int_{-1}^{\eta_0} p(\eta, E)\, d\eta$$

By definition, $P(\eta, E)$ is an increasing function of η, positive and smaller than or equal to unity. To the random variable $H \in (-1, +1)$ we will associate the number $\Xi_3 = P(H, E)$ $[\Xi_3 \in (0,1)]$. Since the correspondence H, Ξ is biunivocal, we can write:

$$Pr[\Xi_3 \in (\xi_3, \xi_3 + d\xi_3)] = Pr[H \in (\eta, \eta + d\eta)]$$

where
$$\begin{cases} \xi_3 = P(\eta, E) \\ d\xi_3 = dP = p(\eta, E)\, d\eta \end{cases}$$

(for a given E).

We deduce

$$Pr[\Xi_3 \in (\xi_3, \xi_3 + d\xi_3)] = d\xi_3$$

The variable Ξ_3 is hence equidistributed in the interval $(0, 1)$.

Consequently, we will calculate the scattering angle Θ, or more precisely its cosine η, in the center of inertia system, when there is anisotropy, by means of the implicit relation:

$$\xi_3 = P(\eta, E)$$

where ξ_3 is a random number equidistributed in the interval $(0, 1)$.

Having tabulated the function $P = P(\eta, E)$ for a discrete sequence of values of E and of η, we can in general determine the η corresponding to a given pair (ξ_3, E) by double interpolation on E and on η. With the notations of the diagram:

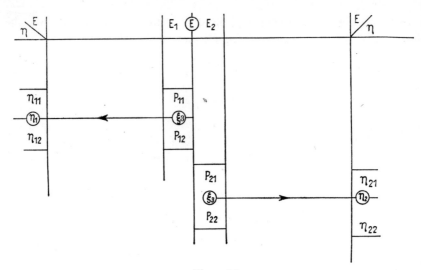

Figure 26

we get successively:

$$\eta_1 = \eta_{11} \frac{\xi_3 - P_{12}}{P_{11} - P_{12}} + \eta_{12} \frac{\xi_3 - P_{11}}{P_{12} - P_{11}}$$

$$\eta_2 = \eta_{21} \frac{\xi_3 - P_{22}}{P_{21} - P_{22}} + \eta_{22} \frac{\xi_3 - P_{21}}{P_{22} - P_{21}}$$

$$\eta = \eta_1 \frac{E - E_2}{E_1 - E_2} + \eta_2 \frac{E - E_1}{E_2 - E_1}$$

whence finally:

$$\cos \Theta = \eta(\xi_3, E)$$

The curve $\eta(\xi_3, E)$ will deviate all the more from the line $2\xi_3 - 1$ as the anisotropy is more pronounced (and in fact, for oxygen, the scattering is more pronounced in front).

We must still determine the variation $\dfrac{E_1}{E_0}$ of the kinetic energy and the angle of deflection θ of the neutron in the laboratory fixed-axes system.

Let $\vec{e}$ be the driving velocity of the moving trihedral. From the conservation of momentum theorem, $\vec{e}$ does not vary during the collision:

$$\vec{e}_0 = \vec{e}_1 = \vec{e}$$

Hence, denoting by $\vec{v}$ and $\vec{V}$ the respective velocities of the neutron and the target with respect to the stationary trihedral:

$$\vec{v}_0 = \vec{e} + \vec{u}_0 \quad \vec{V}_0 = \vec{e} + \vec{U}_0$$

$$\vec{v}_1 = \vec{e} + \vec{u}_1 \quad \vec{V}_1 = \vec{e} + \vec{U}_1$$

In general, we calculate $\vec{e}$, neglecting the velocity $\vec{V}_0$ of the target nucleus before the collision (thermal velocity)*.

$$\vec{e} \simeq -\vec{U}_0 = U\vec{\Omega}_0$$

Now, substituting in the preceding formulas, we get

$$\begin{cases} \vec{v}_0 = (A + 1)\, U\vec{\Omega}_0 \\ \vec{v}_1 = (A\vec{\Omega}_1 + \vec{\Omega}_0)\, U \end{cases}$$

We immediately deduce the variation of the kinetic energy of the neutron:

$$\frac{E_1}{E_0} = \frac{v_1^2}{v_0^2} = \frac{1 + 2A\cos\Theta + A^2}{(1 + A)^2}$$

The angle of deflection θ is the angle between the vectors $\vec{v}_0$ and $\vec{v}_1$. Hence we also have:

$$\cos\theta = \frac{\vec{v}_0 \cdot \vec{v}_1}{\sqrt{v_0^2 \cdot v_1^2}}$$

Hence:

$$\cos\theta = \frac{1 + A\cos\Theta}{\sqrt{1 + 2A\cos\Theta + A^2}}$$

It is not useless to express $\cos\theta$ directly in terms of the energy. Referring to the next to last formula, we get successively:

$$(1 + 2A\cos\Theta + A^2)^{\frac{1}{2}} = (1 + A) \cdot \sqrt{\frac{E_1}{E_0}}$$

$$1 + A\cos\Theta = \frac{1}{2}(1 + A)^2 \frac{E_1}{E_0} - \frac{1}{2}(A^2 - 1)$$

---

* Since the density of the neutron current is very much lower than that of the water, we assume, without any great error, that a nucleus of O or of H is hit at most once by a neutron.

11 Lattes (3006)

whence

$$\cos \theta = \frac{1}{2}(A + 1)\sqrt{\frac{E_1}{E_0}} - \frac{1}{2}(A - 1)\sqrt{\frac{E_0}{E_1}}$$

The preceding formulas are general. We will now apply them to the study of the slowing-down and of the scattering of the neutrons in the water at the ith collision.

a) *Elastic collision with a nucleus of* H

$A = 1$ We get successively:

$$\frac{E_i}{E_{i-1}} = \frac{1 + \cos \Theta}{2}$$

$\cos \Theta = 2\xi_3 - 1$

$$\frac{E_i}{E_{i-1}} = \xi_3$$

$$\cos \theta = \sqrt{\xi_3}$$

The last formula shows that, for an observer connected with the laboratory, the scattering on the hydrogen always occurs forwards. This result must be connected with the fact that

$$\theta = \frac{\Theta}{2}$$

b) *Elastic collision with a nucleus of* O

$A = 16$. The hypothesis $A = \infty$ is a sufficient approximation for the problem under consideration. It allows us to simplify a little the treatment of the collisions with the oxygen. Indeed, the center of inertia is fixed so that there is no longer any difference between central frame of reference and fixed frame of reference. In particular, except for the resonances, the scattering is isotropic with respect to the fixed trihedral.

Hence, passing to the limit

$$\begin{cases} E_i = E_{i-1} \\ \cos \theta = \cos \Theta \end{cases}$$

The probability law of $\theta = \Theta$ depends in fact on E_{i-1}.

As we have seen, three cases occur:

$E_{i-1} > 0.897$ Mev   The scattering is not isotropic. We calculate $\cos \theta$ by double interpolation in tables of numerical values:

$$\cos \theta = \eta(\xi_3, E_{i-1})$$

$0.347 \text{ Mev} < E_{i-1} < 0.544 \text{ Mev}$

The scattering is not isotropic. We have again:

$$\cos \theta = \eta(\xi_3, E_{i-1})$$

$E_{i-1} < 0.897 \text{ Mev}$ The scattering is isotropic for all E_{i-1} not in the preceding band. We simply get:

$$\cos \theta = 2\xi_{3-1}$$

The last formula shows that in the case of oxygen retroscattering is equally as probable as forward scattering. The collisions with oxygen reduce actually to a simple scattering since, with the required approximation, there is no slowing down.

Let $\vec{\omega}_0$ and $\vec{\omega}_1$ be respectively the directions of the neutron before and after the collision. The angle of deflection $\theta = (\vec{\omega}_0, \vec{\omega}_1)$ has been calculated above for two probability laws. To complete the determination of $\vec{\omega}_1$, we must choose a direction on the cone of axis ω_0 and semi-angle θ at the vertex.

The target nucleus and the neutron present circular symmetry at the energies under consideration, so that there is no preferred direction, and we can hence determine the azimuth φ of the plane $\vec{\omega}_0, \vec{\omega}_1$ by choosing an angle at random between 0 and 2π. In practice we set

$$\varphi = 2\pi\xi_4$$

where ξ_4 is a random number equidistributed in the interval $[0, 1]$.

DIRECTION COSINES OF THE PATH

Let $Oxyz$ be the laboratory reference trihedral, with basis:

$$\vec{x}, \vec{y}, \vec{z}$$

$$\vec{\omega}_0 = \alpha_0 \vec{x} + \beta_0 \vec{y} + \gamma_0 \vec{z}$$

$$\vec{\omega}_1 = \alpha_1 \vec{x} + \beta_1 \vec{y} + \gamma_0 \vec{z}$$

First case   The vectors $\vec{z}$ and $\vec{\omega}_0$ form a plane. We are sure of this if $\gamma_0 = 1$. However, in order to avoid indeterminations of a purely numerical nature, on the electronic machines, we agree to keep far away from this limit value. And to do this we distinguish between the two cases

$$\gamma_0 < \tfrac{1}{2} \quad \text{and} \quad \gamma_0 > \tfrac{1}{2}$$

11*

Consider the vectors $\vec{z}$ and $\vec{\omega}_0$ with the same origin. The equatorial planes cut along a direction of unitary vector k normal to the plane $\vec{z}$, $\vec{\omega}_0$ and such that $(\vec{k}, \vec{\omega}_0, \vec{z})$ 0. Let $\vec{i}$ and $\vec{j}$ be the vectors in the plane $\vec{z}$, $\vec{\omega}_0$ defined by the relations

$$\vec{i} = \vec{k} \wedge \vec{\omega}_0$$

$$\vec{j} = \vec{z} \wedge \vec{k}$$

$\vec{i}$ and $\vec{j}$ belong respectively to the equatorial planes of $\vec{\omega}_0$ and $\vec{z}$. We finally denote by $\vec{l}$ the unitary vector of the intersection of the semi-plane $\vec{\omega}_0$, $\vec{\omega}_1$ with the plane $\vec{i}, \vec{k}$. From above, $\vec{\omega}_1$ is defined by the two angles θ and φ:

$$\begin{cases} \theta = (\vec{\omega}_0, \vec{\omega}_1) & 0 \leq \theta \leq \pi \\ \varphi = (\vec{i}, \vec{l}) & 0 \leq \varphi \leq 2\pi \end{cases}$$

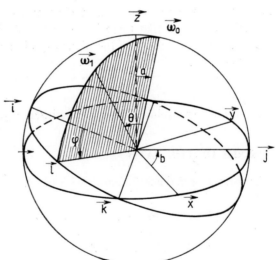

Figure 27

In the reference trihedral $\vec{i}, \vec{k}, \vec{\omega}_0$, we have

$$\vec{\omega}_1 = \sin\theta \cos\varphi\, \vec{i} + \sin\theta \sin\varphi\, \vec{k} + \cos\theta\, \vec{\omega}_0$$

On the other hand, denoting respectively by $a \in (0, \pi)$ and $b \in (0, 2\pi)$ the colatitude and the longitude of $\vec{\omega}_0$ on the unit sphere with axis $0z$, we get successively:

$$\vec{k} = \vec{x} \sin b - \vec{y} \cos b$$

$$\vec{i} = -\vec{j} \cos a + \vec{z} \sin a$$

$$\vec{j} = \vec{x} \cos b + \vec{y} \sin b$$

that is also

$$\vec{i} = -\vec{x} \cos a \cos b - \vec{y} \cos a \sin b + \vec{z} \sin a$$

Substituting this expression for $\vec{i}$ and that for $\vec{k}$ in the formula giving $\vec{\omega}_1$, we get:

$$\vec{\omega}_1 = \vec{x} \sin \theta \, [- \cos a \cos b \cos \varphi + \sin b \sin \varphi)$$

$$+ \vec{y} \sin \theta \, [- \cos a \sin b \cos \varphi - \cos b \sin \varphi]$$

$$+ \vec{z} \sin \theta \sin a \cos \varphi + \omega_0 \cos \theta$$

This formula expresses $\vec{\omega}_1$ in terms of $\vec{\omega}_0$, θ and φ.

It introduces two auxilliary variables a and b, functions of α_0, β_0 and γ_0 by means of the relations

$$\sin a \cos b = \alpha_0$$

$$\sin a \sin b = \beta_0$$

$$\cos a = \gamma_0$$

We deduce readily, since $a \in (0, \pi)$ and $\gamma_0 \neq 1$:

$$\cos b = + \frac{\alpha_0}{\sqrt{1 - \gamma_0^2}}$$

$$\sin b = + \frac{\beta_0}{\sqrt{1 - \gamma_0^2}}$$

$$\sin a = + \sqrt{1 - \gamma_0^2}$$

$$\cos a = \gamma_0$$

Substituting in the expression for $\vec{\omega}_1$ we finally get

First case $\gamma_0 < \tfrac{1}{2}$

$$\vec{\omega}_1 = \vec{x}\left\{ \frac{\sin\theta}{\sqrt{1-\gamma_0^2}}(-\alpha_0\gamma_0\cos\varphi + \beta_0\sin\varphi) + \alpha_0\cos\theta \right\}$$

$$+ \vec{y}\left\{ \frac{\sin\theta}{\sqrt{1-\gamma_0^2}}(-\beta_0\gamma_0\cos\varphi - \alpha_0\sin\varphi) + \beta_0\cos\theta \right\}$$

$$+ \vec{z}\left\{ \sin\theta\cdot\sqrt{1-\gamma_0^2}\cos\varphi + \gamma_0\cos\theta \right\}$$

Direction of the neutron after the ith collision when $\gamma_{i-1} < \tfrac{1}{2}$

The probability laws of the variables $\theta \in (0, \pi)$ and $\varphi \in (0, 2\pi)$ are known:

$$\cos\theta = f(\xi_3)$$

$$\varphi = 2\pi\xi_4$$

Denoting respectively by $\vec{\omega}_{i-1}(\alpha_{i-1}, \beta_{i-1}, \gamma_{i-1})$ and $\vec{\omega}_i(\alpha_i, \beta_i, \gamma_i)$ the directions of the neutron before and after the ith collision, we get:

$$\alpha_i = \frac{\sqrt{1-f^2(\xi_3)}}{\sqrt{\alpha_{i-1}^2 + \beta_{i-1}^2}}[-\alpha_{i-1}\gamma_{i-1}\cos 2\pi\xi_4 + \beta_{i-1}\sin 2\pi\xi_4] + \alpha_{i-1}f(\xi_3)$$

$$\beta_i = \frac{\sqrt{1-f^2(\xi_3)}}{\sqrt{\alpha_{i-1}^2 + \beta_{i-1}^2}}[-\beta_{i-1}\gamma_{i-1}\cos 2\pi\xi_4 - \alpha_{i-1}\sin 2\pi\xi_4] + \beta_{i-1}f(\xi_3)$$

$$\gamma_i = \sqrt{1-f^2(\xi_3)}\cdot\sqrt{\alpha_{i-1}^2 + \beta_{i-1}^2}\ \cos 2\pi\xi_4 + \gamma_{i-1}f(\xi_3)$$

Second case

$$\gamma_0 > \tfrac{1}{2}$$

In this case the vectors $\vec{y}$ and $\vec{\omega}_0$ certainly form a plane. The calculation is in all ways analogous to the preceding one.

The diagram is the same, except that $\vec{z}$ is replaced by $\vec{y}$, $\vec{x}$ by $\vec{z}$, $\vec{y}$ by $\vec{x}$.

We can hence go from one group of formulas to the other by performing the double circular permutation:

$$\gamma_0 \to \beta_0 \to \alpha_0 \to \gamma_0$$

$$\vec{z} \to \vec{y} \to \vec{x} \to \vec{z}$$

whence:

$$\vec{\omega}_1 = \vec{x} \left\{ \frac{\sin \theta}{\sqrt{1 - \beta_0^2}} (-\alpha_0 \beta_0 \cos \varphi - \gamma_0 \sin \varphi) + \alpha_0 \cos \theta \right\}$$

$$+ \vec{y} \left\{ \sin \theta \cdot \sqrt{1 - \beta_0^2} \cos \varphi + \beta_0 \cos \theta \right\}$$

$$+ \vec{z} \left\{ \frac{\sin \theta}{\sqrt{1 - \beta_0^2}} (-\beta_0 \gamma_0 \cos \varphi + \alpha_0 \sin \varphi) + \gamma_0 \cos \theta \right\}$$

Direction of the neutron after the ith collision when $\gamma_{i-1} > \frac{1}{2}$

The probability laws of the variables $\theta \in (0, \pi)$ and $\varphi \in (0, 2\pi)$ are known

$$\cos \theta = f(\xi_3)$$

$$\varphi = 2\pi \xi_4$$

Denoting respectively by $\vec{\omega}_{i-1}(\alpha_{i-1}, \beta_{i-1}, \gamma_{i-1})$ and $\vec{\omega}_i(\alpha_i, \beta_i, \gamma_i)$ the directions of the neutron before and after the *i*th collision we get:

$$\alpha_i = \frac{\sqrt{1 - f^2(\xi_3)}}{\sqrt{\alpha_{i-1}^2 + \gamma_{i-1}^2}} (\alpha_{i-1} \beta_{i-1} \cos 2\pi \xi_4 - \gamma_{i-1} \sin 2\pi \xi_4) + \alpha_{i-1} f(\xi_3)$$

$$\beta_i = \sqrt{1 - f^2(\xi_3)} \ \sqrt{\gamma_{i-1}^2 + \alpha_{i-1}^2} \ \cos 2\pi \ \xi_4 + \beta_{i-1} f(\xi_3)$$

$$\gamma_i = \frac{\sqrt{1 - f^2(\xi_3)}}{\sqrt{\alpha_{i-1}^2 + \gamma_{i-1}^2}} (-\beta_{i-1} \gamma_{i-1} \cos 2\pi \xi_4 + \alpha_{i-1} \sin 2\pi \xi_4) + \gamma_{i-1} f(\xi_3)$$

Solution of Differential, Difference and Partial Differential Equations by the Monte Carlo Method

THE GAMBLER'S RUIN PROBLEM

Two players G and G' with the sums g and g', respectively, decide to play against one another in a game J until complete ruin of one of them; J consists of the repetition of the same play, with constant stakes h^* which G and G' can win with the respective probabilities p and $q = 1 - p$.

When $p < q$, it might nevertheless be in G's interest to play if his capital is very much larger than that of his opponent and if his stake is not too small.

Conversely, when $p > q$, it might nevertheless be in G's interest to withdraw if his capital is small and if the stakes are high.

The evolution of the capital x of G can be followed on an axis $0x$ between the points A, $x = 0$ (ruin of G) and B, $x = g + g' = b$ (ruin of G'), where the representative point $M(x)$ moves by jumps of amplitude $\pm h$ with the probabilities:

$$\begin{cases} p = \text{Prob } [x \to x + h] \\ q = \text{Prob } [x \to x - h] \end{cases}$$

$$0 < x < b$$

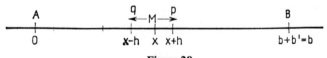

Figure 28

---

* G or G' are eliminated when the capital is smaller than h, so that we can always suppose that the stake is a common divisor of g and g'.

The game is more or less long. *A priori* it might even extend indefinitely. Before passing to the limit, we must examine what happens in a given time.

After m plays, G wins or loses unless there is indecision. Let $v_m(x)$, $u_m(x)$, $w_m(x)$ be the corresponding probabilities:

$$v_m(x) + u_m(x) + w_m(x) = 1$$

Victory or defeat occur during the 0th, or the 1st, or the 2nd, $\cdots$, or the mth play*. Consequently $v_m(x)$ and $u_m(x)$ are non-decreasing functions of m. Also $w_m(x)$ is a non-increasing function of m. The sequences $v_m(x)$, $u_m(x)$ and $w_m(x)$, bounded and monotonic, have limits $v(x)$, $u(x)$ and $w(x)$ when m increases indefinitely. We will calculate these limits. We will show in particular that $w(x) \equiv 0$, that is to say that the game will certainly end by the ruin of one of the players.

In practice, it suffices to calculate $v(x) \equiv v(x, p, q)$ since, by symmetry:

$$u(x) \equiv v(b - x, q, p)$$
$$w(x) \equiv 1 - [v(x) + u(x)]$$

More rigorously, we will start by establishing the general equation in a game restricted to m plays and then we will pass to the limit.

Equation in $v_m(x)$

Let x be the initial capital of G.

When $x = 0$ or $x = b$, there is no real game since G or G' are ruined at the start. We then have

$$\begin{cases} v_m(0) = 0 \\ v_m(b) = 1 \end{cases} \quad m = 0, 1, 2, \cdots$$

If $0 < x < b$, at least one play is necessary in order to find a winner. After this play, the capital of G is $x + h$ or $x - h$ and he still has at most $m - 1$ plays to play in order to beat G'. If G wins, a step towards his victory will necessarily be

$$\{x \to x + h \to b\} \quad \text{or} \quad \{x \to x - h \to o\}$$

with respective probability $pv_{m-1}(x + h)$ and $qv_{m-1}(x - h)$. It follows that:

$$v_m(x) = pv_{m-1}(x + h) + qv_{m-1}(x - h)$$

---

* If $x = b$ G wins by forfeit (play 0)
  If $x = 0$ G loses by withdrawal (play 0).

This is a difference equation with two variables x and m. The limit conditions are:

$$\begin{cases} v_m(0) = 0 \\ v_m(1) = 1 \end{cases} \quad \text{for} \quad m = 0, 1, 2 \cdots$$

and $\qquad v_0(x) = 0 \quad \text{for} \quad 0 < x < b$

Equation in $v(x)$

We now agree to extend the game beyond the mth play, if no winner has yet been found. G's chances are increasing. Supposing the players are infinitely patient, they can however not exceed, at the limit, the rate $v(x) = \lim\limits_{m \to \infty} v_m(x)$.

Equating the limit values of each of the two members of the equation in $v_m(x)$ we get:

$$v(x) = pv(x + h) + qv(x - h)$$
$$\begin{cases} v(0) = 0 \\ v(1) = 1 \end{cases}$$

a difference equation defining $v(x)$.

In the present case $[p + q = 1]$ it assumes the remarkable form:

$$p[v(x + h) - v(x)] = q[v(x) - v(x - h)]$$

which shows that $v(x)$ is monotonic and, more precisely, increasing, if we take the initial conditions into account.

Calculation of $v(x)$ Set $b = Nh$, where N is a positive integer.

The difference equation allows us to calculate the value of $v(x)$ at the $N + 1$ points of the net $x = nh, n = 0, 1, 2, ..., N$.

Passing from the notations of functions to those of sequences, we write:

$$v(x) = v(nh) = v_n$$

The sequence v_n will be defined by the system:

$$\begin{cases} v_{n+1} - v_n = \dfrac{q}{p} (v_n - v_{n-1}) & n = 1, 2, ..., N - 1 \\[2mm] v_0 = 0 \\[1mm] v_N = 1 \end{cases}$$

which contains indeed $N - 1$ equations and two initial conditions to determine the $N + 1$ values $v_0, v_1, ..., v_N$.

We write explicitly the $n - 1$ first equations of the system:

$$v_n - v_{n-1} = \frac{q}{p}(v_{n-1} - v_{n-2})$$

$$v_{n-1} - v_{n-2} = \frac{q}{p}(v_{n-2} - v_{n-3})$$

$$\cdots \cdots \cdots \cdots \cdots \cdots \cdots$$

$$v_2 - v_1 = \frac{q}{p}(v_1 - v_0)$$

Multiplying side by side and eliminating the common factors as we go along*, we get:

$$v_n - v_{n-1} = \left(\frac{q}{p}\right)^{n-1} v_1 \quad n = 1, 2, ..., N$$

taking the condition $v_0 = 0$ into account.

We write again the n first equations of the above sequence:

$$v_n - v_{n-1} = \left(\frac{q}{p}\right)^{n-1} \cdot v_1$$

$$v_{n-1} - v_{n-2} = \left(\frac{q}{p}\right)^{n-2} \cdot v_1$$

$$\cdots \cdots \cdots \cdots \cdots \cdots \cdots$$

$$v_1 - v_0 = 1 \cdot v_1$$

Adding corresponding sides we get, after $n - 1$ reductions

$$v_n = \left[1 + \frac{q}{p} + \left(\frac{q}{p}\right)^2 + \cdots + \left(\frac{q}{p}\right)^{n-1}\right] v_1$$

We determine the value of v_1 by taking the second initial condition $v_N = 1$ into account:

$$1 = \left[1 + \frac{q}{p} + \left(\frac{q}{p}\right)^2 + \cdots + \left(\frac{q}{p}\right)^{N-1}\right] v_1$$

---

* These factors are not zero since the sequence is increasing.

which shows that v_1 is not zero. We deduce:

$$v_n = \frac{1 + \dfrac{q}{p} + \left(\dfrac{q}{p}\right)^2 + \cdots + \left(\dfrac{q}{p}\right)^{n-1}}{1 + \dfrac{q}{p} + \left(\dfrac{q}{p}\right)^2 + \cdots + \left(\dfrac{q}{p}\right)^{N-1}}$$

v_n is thus the ratio of two geometric series with argument $\dfrac{p}{q}$.
If $p = q = \tfrac{1}{2}$ we get simply

$$v_n = \frac{n}{N} \quad \text{that is} \quad v(x) = \frac{x}{b}$$

For constant sum, the probability of ruining G' is proportional to the capital of G. The game is fair only if the capitals are equal.

If $p \neq q$ we get, multiplying above and below by $1 - \dfrac{p}{q}$:

$$v_n = \frac{1 - \left(\dfrac{q}{p}\right)^n}{1 - \left(\dfrac{q}{p}\right)^N} \quad \text{that is} \quad v(x) = \frac{1 - \left(\dfrac{q}{p}\right)^{x/h}}{1 - \left(\dfrac{q}{p}\right)^{b/h}}$$

For constant b, the probability of ruining G' is an increasing exponential function of the capital x of G. For small x, the growth is rapid when $p > q$, slow when $p < q$. The reverse holds for high x.

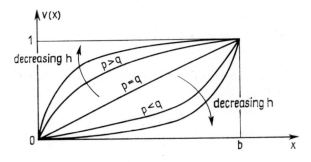

Figure 29

G's chances depend also on the stake h. Letting it vary we get a series of curves $v(x)$ which are located respectively above and below the straight line

$$v(x) = \frac{x}{b}$$

depending on whether $p > q$ or $p < q$.

In the upper triangle $v(x)$ is a decreasing function of the stake. In the lower triangle, it is, on the contrary, an increasing function*.

Consequently, when the plays are in his favour $(p > q)$ it is in G's interest to decrease the stake and this all the more so as his capital is smaller. On the contrary, when G is inclined to lose the plays $(p < q)$ he can hope to ruin his opponent only by starting with a large sum and playing with high stakes.

Probability of the ruin of G, $u(x)$

It is deduced from $v(x, p, q)$ by the permutation:

$$u(x) = v(b - x, q, p)$$

If $p \neq q$, we get

$$u(x) = \frac{1 - \left(\dfrac{p}{q}\right)^{b/h} \left(\dfrac{p}{q}\right)^{-x/h}}{1 - \left(\dfrac{p}{q}\right)^{b/h}} = \frac{\left(\dfrac{q}{p}\right)^{x/h} - \left(\dfrac{q}{p}\right)^{b/h}}{1 - \left(\dfrac{q}{p}\right)^{b/h}}$$

We immediately deduce the identity:

$$v(x) + u(x) \equiv 1$$

Consequently $w(x) \equiv 0$. The game will thus always have a winner, no matter the p, q, h, x and b.

---

* These results can readily be proved by noting that $v(x)$ depends in fact only on the parameters r and θ:

$$\left. \begin{array}{ll} r \equiv \left(\dfrac{q}{p}\right)^{b/h} & 0 < r \\[3mm] \theta \equiv \dfrac{x}{b} & 0 < \theta < 1 \end{array} \right\} \quad v(\theta r) \equiv \frac{1 - r^\theta}{1 - r}.$$

Now $v(r)$ is a decreasing function. Consequently $v(h)$ varies inversely to $r(h)$.

If $p > q \rightarrow 0 < r < \dfrac{q}{p} < 1$, $v(h)$ is a decreasing function because $r(h)$ is an increasing function.

If $p < q \rightarrow 1 < \dfrac{q}{p} < r$, $v(h)$ is an increasing function because $r(h)$ is a decreasing function.

If $p = q$, we have:

$$u(x) = 1 - \frac{x}{b}$$

whence $\qquad\qquad u(x) + v(x) \equiv 1 \quad \text{and} \quad w(x) \equiv 0$

the same conclusion as above: the capital of one of the gamblers will always finally vanish if the game lasts long enough.

Remark   From the note (1), the necessary and sufficient condition for two homothetic games $\left[\theta = \theta_0, \text{ that is} \frac{x}{b} = \frac{x_0}{b_0} \right]$ to correspond to the same probability law is that $r = r_0$, that is

$$\left(\frac{q}{p} \right)^{b/h} = \left(\frac{q_0}{p_0} \right)^{b_0/h_0}$$

a relation between the four parameters p, q, h and b. In particular, the games with low stakes, h, necessarily correspond to nearly equal plays:

$$p \simeq q \simeq \tfrac{1}{2}$$

since $\qquad\qquad \frac{b}{h} \cdot \log \frac{q}{p} = K_0$

$$\log \frac{q}{p} = \frac{K_0}{b} \cdot h$$

$$\frac{q}{p} = 1 + \frac{K_0}{b} h + 0(h)$$

where K_0 is a constant number.

This result must be connected with the fact that a *finite* difference $p - q$ in a small-gain game (very small h) would lead with certainty to the victory of G or G' depending on whether $p > q$ or $q < p$.

This being so, when $h \to 0$, the net $N = b/h$ tightens. At the limit, the sequence v_n ($n = 0, 1, 2, \cdots, N$) finally becomes identical to a function $v(x)$ of the continuous variable $x \in (0, b)$.

By definition of the similitude, $v(x)$ performs the smoothing of the $N_0 + 1$ points v_{n0} of the discrete, primitive network. It is hence possible to represent the probabilities in a discontinuous problem by means of a continuous function. In particular, there exists a differential equation equivalent to the limit of the difference equation of the concrete problem we considered. This is what we are now going to study.

Equivalent differential equation

The difference equation:

$$v(x) = pv(x + h) + qv(x - h)$$

is of the second order since it introduces three consecutive points. It can hence be expressed in terms of the first and second differences $\Delta v(x)$ and $\Delta^2 v(x)$:

$$\Delta v(x) \equiv \frac{v(x + h) - v(x)}{h}$$

$$\Delta^2 v(x) \equiv \frac{\Delta v(x) - \Delta v(x - h)}{h} = \frac{1}{h^2} [v(x + h) - 2v(x) + v(x - h)]$$

We have indeed, successively:

$$p[v(x + h) - v(x)] = q[v(x) - v(x - h)]$$

$$p \Delta v(x) = q \Delta v(x - h)$$

$$(p - q) \Delta v(x) = q[\Delta v(x - h) - \Delta v(x)]$$

$$(p - q) \Delta v(x) = -qh \Delta^2 v(x)$$

that is finally:

$$(I) \qquad \begin{cases} \Delta^2 v(x) + \dfrac{p - q}{qh} \Delta v(x) = 0 \\[2mm] x = nh \quad b = Nh \\[2mm] v(0) = 0 \quad v(b) = 1 \\[2mm] n = 1, 2, ..., N - 1 \end{cases}$$

The equation (I) defines the $v_n \equiv v(nh)$ in terms of the parameters h and $2\alpha \equiv \dfrac{p - q}{qh}$. It has already been solved by direct computation. We can make the differential equation (II) correspond to it

$$(II) \qquad \begin{cases} \dfrac{d^2 v}{dx^2} + 2\beta \dfrac{dv}{dx} = 0 \\[2mm] v(0) = 0 \\[2mm] v(b) = 1 \end{cases}$$

whose solution $v(n)$ coincides with the sequence v_n at all points of the net.

Calculation shows that this is so when

$$2\beta \equiv \frac{1}{h}\log\frac{p}{q}$$

Indeed, any solution of (II) is a linear combination of the two particular solutions 1 and $e^{-2\beta x}$.

We hence get successively:

$$v(x) = A + Be^{-2\beta x}$$

$$\Delta v(x) = Be^{-2\beta x}\frac{e^{-2\beta h} - 1}{h}$$

$$\Delta v(x - h) = Be^{-2\beta x}\frac{1 - e^{2\beta h}}{h}$$

$$\Delta^2 v(x) = Be^{-2\beta x}\frac{e^{2\beta h} + e^{-2\beta h} - 2}{h^2}$$

$$= Be^{-2\beta x}\frac{(1 - e^{2\beta h})(e^{-2\beta h} - 1)}{h^2}$$

Comparing the first and second differences, we verify that any solution of (II) satisfies the difference equation:

$$\Delta^2 v(x) + \frac{e^{2\beta h} - 1}{h}\Delta v(x) = 0$$

which coincides with (I) if $2\alpha = \dfrac{e^{2\beta h} - 1}{h}$, that is

$$2\beta = \frac{1}{h}\log(1 + 2\alpha h) \equiv \frac{1}{h}\log\frac{p}{q}.$$

Remarks When the stakes are low [$|2\alpha h| \ll 1$] β differs very little from α because:

$$2\beta \approx 2\alpha - \frac{2\alpha^2}{2}\cdot h + \frac{2\alpha^3}{3}h^2 - \cdots$$

and we can calculate an approximate value of $v(x)$ from the equation (II')

(II')
$$\left\{ \begin{array}{l} \dfrac{d^2v}{dx^2} + 2x\dfrac{dv}{dx} = 0 \\[2mm] v(0) = 0 \\[1mm] v(b) = 1 \end{array} \right.$$

which is merely the limit form of the difference equation (I), when $h \to 0$ for constant x.

In conclusion, we can calculate the sequence v_n from one of the equations (I), (II) or (II'). Integration of the latter is classical. For instance, (II) readily gives, taking the limit conditions into account:

$$\left\{ \begin{array}{l} v(x) = A + Be^{-2\beta x} \\[1mm] 0 = A + B \\[1mm] 1 = A + Be^{-2\beta b} \end{array} \right.$$

whence
$$v(x) = \dfrac{1 - e^{-2\beta x}}{1 - e^{-2\beta b}}$$

that is again, since
$$2\beta = \dfrac{1}{h}\log\dfrac{p}{q}$$

$$v(x) = \dfrac{1 - \left(\dfrac{q}{p}\right)^{x/h}}{1 - \left(\dfrac{q}{p}\right)^{b/h}}$$

APPLICATION—SOLUTION OF DIFFERENTIAL AND DIFFERENCE EQUATIONS BY THE MONTE CARLO METHOD

1) We have seen that the integral of the differential equation (II) represents the *exact* value of the transfer probability $x \to b$, $x \neq 0$, in a game p, q, h, defined by the condition

$$2\beta = \dfrac{1}{h}\log\dfrac{p}{q}$$

Conversely, if the two opponents G and G' play a large number of times a game p, q, h, satisfying the above condition, the frequency of G's failures will represent a statistical value of the integral $v(x)$ of the limit problem

$$
\begin{cases}
\dfrac{d^2v}{dx^2} + 2\beta\,\dfrac{dv}{dx} = 0 \\[2mm]
v(0) = 0 \\[2mm]
v(b) = 1
\end{cases}
$$

The Monte Carlo method is based precisely on the realisation of this game. We take the step

$$
h = \frac{b}{m} \quad \text{(where } m \text{ is an integer)}
$$

we deduce p and q such that

$$
\begin{cases}
p + q = 1 \\[2mm]
\dfrac{p}{q} = e^{2\beta h}
\end{cases}
$$

that is

$$
p = \frac{e^{2\beta h}}{1 + e^{2\beta h}}, \quad q = \frac{1}{1 + e^{2\beta h}}
$$

We take a starting point x.

We pick a random number $\xi \in [0, 1]$.

If $0 \leq \xi < p$, we move the particle from x to $x + h$

If $p \leq \xi < 1$, we move the particle from x to $x - h$

The game is continued until the particle is in A or B. In the first case, we score $s = 0$; in the second, we score $s = 1$.

We begin the game again n times starting always from the same point s. Say

$$
\tilde{s} = \frac{1}{N} \sum_{i=1}^{N} s_i
$$

is the average of the score s over N plays. $\tilde{s}$ represents an approximate value of the integral $v(x)$, at x. The approximation is all the better as N is larger because

$$
\lim_{N \to \infty} \tilde{s} = \langle s \rangle
$$

and

$$
\langle s \rangle = 1 \cdot v(x) + 0 \cdot [1 - v(x)] = v(x)
$$

Setting ourselves successively at all the points of the net $x = nh$, $n = 1, 2, \ldots, m - 1$, we finally obtain a step curve giving the integral $v(x)$ in the interval $(0, b)$.

There are three ways of improving the results: increasing m, or N, or m and N.

For constant m, the error varies as $\dfrac{1}{\sqrt{N}}$.

For constant N, the increase of m gives us a better representation of the unknown function in the regions where it varies rapidly. However this result is obtained at the cost of a large increase in the length of the plays because games with low stakes, h (high m), correspond to near draws $p \simeq q \simeq \frac{1}{2}$. We have indeed, for $2\beta h \ll 1$:

$$p = \tfrac{1}{2}(1 + \beta h) + 0(h)$$

$$q = \tfrac{1}{2}(1 - \beta h) + 0(h)$$

that is

$$p - q = h\beta + 0(h)$$

2) The numerical solution of the difference equation (I)

$$\begin{cases} \Delta^2 v(x) + 2\alpha\, \Delta v(x) = 0 \\ b = mh \quad x = nh \quad n = 1, 2, \ldots, m - 1 \\ v(0) = 0 \\ v(b) = 1 \end{cases}$$

makes use of the same methods. We have seen, indeed, that the solution $v(x)$ represents the *exact* value of the transfer probability $x \to b$, $x \neq 0$, in the game p, q, h defined by the condition

$$2\alpha = \frac{p - q}{qh}$$

Conversely if the two opponents G and G' play a large number of times the game p, q, h, satisfying the above condition, the frequency of the failures of G' will represent a statistical evaluation of $v(x)$.

The practical organization of the game will be the same as above, except that the stake h is now given (it is the step of the net). The numbers p and q

12*

are deduced from it by

$$p + q = 1$$

$$\frac{p - q}{q} = 2\alpha h$$

whence

$$\begin{cases} p = \dfrac{1}{2} \dfrac{1 + 2\alpha h}{1 + \alpha h} \\[3mm] q = \dfrac{1}{2} \dfrac{1}{1 + \alpha h} \end{cases}$$

We will note, again, that the low-stakes games allow a good representation of the solution, at the cost, however, of a large increase in the length of the plays. We have again, indeed, for $\alpha h \ll 1$:

$$p = \tfrac{1}{2}(1 + \alpha h) + 0(h)$$

$$q = \tfrac{1}{2}(1 - \alpha h) + 0(h)$$

that is

$$p - q = \alpha h \qquad\qquad + 0(h)$$

Remark 1 Equations (I) and (II) have been solved independently from one another by the Monte Carlo method. The fact that the results are identical when $\alpha \equiv \beta$ and $h \to 0$ confirms the fact that the difference equation gives a very good approximation of the small-step difference equation. Consequently, each time its form is suitable we need only use the Monte Carlo method on the difference equation.

Proceeding in this way, we add a systematic error $\dfrac{\partial v}{\partial h} h$ to the statistical error due to the error in the results, but the first will, in general, be negligible compared with the second.

Remark 2 If we introduce the number of moves m of a play as an additional variable, the development of the game which opposes two opponents until the ruin of one of them satisfies a difference equation of parabolic type. Conversely, we can show that the formalism of the gambler's ruin problem is well suited to the simulation of certain partial differential equations of parabolic type. This has already been verified on the differential equation (II). We will now study two generalizations of this equation whose treatment makes use of the same method.

Solution of differential equations with varying coefficients

We consider second-order equations of type:

$$(\text{III}) \quad \begin{cases} LU(x) \equiv \beta(x)\dfrac{d^2U}{dx^2} + 2\alpha(x)\dfrac{dU}{dx} = 0 \\[2mm] a < x < b \\[2mm] U(a) = A \quad U(b) = B \end{cases}$$

where $\alpha(x)$ and $\beta(x)$ are two functions of x defined and bounded in the interval (a, b) but not necessarily continuous. Moreover, $\beta(x) > 0$ for all $x \in (a, b)$.

The method of finite differences leads to the substitution for the interval (a, b) of a net of steps h, for the differential operator L of a difference operator L_h and for the unknown function $U(x)$ of the function $V(x, h)$ satisfying the equation*:

$$\text{III}_h \quad \begin{cases} L_h V(x, h) \equiv \beta(x)\,\Delta^2 V + 2\alpha(x)\,\Delta V = 0 \\[2mm] x = nh \in (a, b) \\[2mm] V(x, h) \equiv A \quad \text{if} \quad x \le a \\[2mm] V(x, h) \equiv B \quad \text{if} \quad x \ge b \end{cases}$$

This system is generally solved by iteration. We wish here to calculate $V(x, h)$ by assimilating it with the mean value of a random variable which is suitably defined. In order to keep the same notations, we will denote henceforth by $V(x)$ the solution of (III_h)†.

Making the differences explicit, (III_h) can also be written:

$$\beta(x)\,[V(x + h) + V(x - h) - 2V(x)] + 2\alpha(x)\,h\,[V(x + h) - V(x)] = 0$$

whence we go to

$$V(x) = p(x)\,V(x + h) + q(x)\,V(x - h)$$

where

$$\begin{cases} p(x) \equiv \dfrac{1}{D(x)}\,[\beta(x) + 2h\alpha(x)] \\[3mm] q(x) \equiv \dfrac{1}{D(x)}\,\beta(x) \\[3mm] D(x) \equiv 2\beta(x) + 2h\alpha(x) \end{cases}$$

---

* Assuming the existence of the derivatives and evaluating the difference $L_h V(x) - LV(x)$ for an arbitrary function $V(x)$, we readily show that $V(x, h)$ is very close to $U(x)$ when h is small.

† Even though it strictly speaking depends on the step.

When $\alpha(x) < 0$, we can always choose h sufficiently small so that $2h|\alpha(x)| < \beta(x)$ since $\alpha(x)$ is bounded in the interval (a, b). The quantities $p(x)$ and $q(x)$ can hence be considered as positive. They also satisfy the identity

$$p(x) + q(x) \equiv 1$$

which suggests the existence of a statistical model where p and q would play the role of transition probabilities $x \rightarrow x + h$ and $x \rightarrow x - h$.

In order to carry out further the analogy with equation (I), we would have to identify $V(x)$ with the mathematical hope $\langle s \rangle$ of a random variable s assuming the value $V(x') = A$ when a path originating from x ends first at $x' \leq a$ or assuming the value $V(x') = B$ when the possibility $x \rightarrow x' \geq b$ occurs first.

We will verify that this is indeed so.

The game p, q, h develops by successive stages of amplitude h, which lead the particle from its starting point x to a point x'. The path ends when the particle reaches or crosses the extremities of the interval (a, b) (we can say that it is absorbed). Many paths are possible. Some of them are very long. In order to classify them it is convenient to examine first those which have no more than m steps. There are three types of these corresponding to the three possibilities: the particle is absorbed at A, the particle is absorbed at B, the particle survives the mth transition. We will denote by $v_m(x, a)$, $v_m(x, b)$ and $v_m(x)$ their respective probabilities:

$$v_m(x, a) + v_m(x, b) + v_m(x) = 1$$

The two first are non-decreasing functions of m*. Also, the last is a non-increasing function of m. The sequences $v_m(x, a)$, $v_m(x, b)$ and $v_m(x)$, which are bounded and monotonic, hence have limits $v(x, a)$, $v(x, b)$ and $v(x)$ when m increases indefinitely. Moreover, $p(x)$ and $q(x)$ do not vanish in the interval (a, b) (intuitively, a particle always ends up by being absorbed). Hence, we must have $V(x) \equiv 0$†.

---

* Let $\varphi_{m+1}(x, a)$ and $\varphi_{m+1}(x, b)$ be respectively the fractions of particles which survive the mth transition and are absorbed during the $(m + 1)$st transfer. We have:

$$v_{m+1}(x, a) = v_m(x, a) + v_m(x)\,\varphi_{m+1}(x, a) \geq v_m(x, a)$$
$$v_{m+1}(x, b) = v_m(x, b) + v_m(x)\,\varphi_{m+1}(x, b) \geq v_m(x, b).$$

† This result has been proved earlier (cf. The gambler's ruin problem) in the particular case where the transfer probabilities p and q are constant. We will assume it here, without proof.

We must now establish the relation between the development of the game and the calculation of the integral of the equation under consideration.

The probabilities $v_m(x, a)$ and $v_m(x, b)$ satisfy the equations (cf. the gambler's ruin problem):

$$
\begin{cases}
a < x < b \\
v_m(x, a) = p(x) \cdot v_{m-1}(x + h, a) + q(x) \cdot v_{m-1}(x - h, a) \\
v_0(x, a) \equiv v_0(x, b) \equiv 0 \quad \text{if} \quad a < x < b \\
v_m(x, a) \equiv 1 \quad \text{if} \quad x \leq a \\
v_m(x, a) \equiv 0 \quad \text{if} \quad x \geq b
\end{cases}
$$

$$
\begin{cases}
a < x < b \\
v_m(x, b) = p(x) \, v_{m-1}(x + h, b) + q(x) \, v_{m-1}(x - h, b) \\
v_0(x, a) \equiv v_0(x, b) \equiv 0 \quad \text{if} \quad a < x < b \\
v_m(x, b) \equiv 0 \quad \text{if} \quad x \leq a \\
v_m(x, b) \equiv 1 \quad \text{if} \quad x \geq b
\end{cases}
$$

Denote by $V_m(x)$ the mean value of a random variable s (the score) which assumes the value A, B or 0 depending on whether a particle originating from x is absorbed at A or B, at the latest at the mth transition, or survives this transition:

$$
V_m(x) \equiv \langle s \rangle_m = A v_m(x, a) + B v_m(x, b)
$$

By substitution, we verify that $V_m(x)$ satisfies the same difference equation as the absorption probabilities. We have, indeed:

$$
\begin{cases}
a < x < b \\
V_m(x) = p(x) \, V_{m-1}(x + h) + q(x) \, V_{m-1}(x - h) \\
V_0(x) \equiv 0 \quad \text{if} \quad a < x < b \\
V_m(x) \equiv A \quad \text{if} \quad x \leq a \\
V_m(x) \equiv B \quad \text{if} \quad x \geq b
\end{cases}
$$

From this result we can deduce several consequences:

1) The organization of a transport game $p(x)$, $q(x)$, h, restricted to a given number of transitions allows us to solve numerically the difference equations with two variables x and m of the preceding type, when they themselves are derived from a more general equation, with second-order partial derivatives (see below).

2) From the existence of limits $v_m(x, a) \to v(x, a)$ and $v_m(x, b) \to v(x, b)$ when $m \to \infty$, we deduce the existence of a limit for the function $V_m(x)$ itself, namely $\bar{V}(x)$.

$$\bar{V}(x) \equiv \langle s \rangle = Av(x, a) + Bv(x, b)$$

for all constants A and B. This limit is also that of $V_{m-1}(x)$, and hence it satisfies the equation:

$$\text{IV}_h \quad \begin{cases} \bar{V}(x)\, p(x)\, \bar{V}(x + h) + q(x)\, \bar{V}(x - h) \\[6pt] a < x < b \\[6pt] \bar{V}(x) \equiv A \quad \text{if} \quad x \leq a \\[6pt] \bar{V}(x) \equiv B \quad \text{if} \quad x \leq b \end{cases}$$

We recognize the system (III_h). Consequently, $\bar{V}(x)$ is merely the integral $V(x)$:

$$\lim_{m \to \infty} V_m(x) \equiv \bar{V}(x) \equiv V(x)$$

The transport game, continued until the particle is captured, gives us thus a calculation procedure for differential equations of type (III).

Remark   For small steps, the transition probabilities are quasi-isotropic:

$$p(x) - q(x) = h \cdot \frac{\alpha(x)}{\beta(x)} + 0(h)$$

It follows that each path contains a large number of very short stages. However convergence is guaranteed because no particle can survive.

Solution of parabolic partial differential equations

In the above we were only interested in paths with no more than m transitions. If λ represents the time interval corresponding to any path with amplitude h, the previous description gives the percentage of particles which are or are not captured in the time interval $t = m\lambda^*$. Time is thus naturally

---

* We suppose that there is no pause between two successive transitions.

introduced as a second variable. We set, by definition:

$$v_m(x, a) \equiv v_\lambda(x, t; a)$$

$$v_m(x, b) \equiv v_\lambda(x, t; b) \quad \text{etc.}$$

We will represent similarly the average score $V_m(x) \equiv \langle s \rangle_m$ associated with the behaviour of a large number of particles in the time $t = m\lambda$ by the notation

$$V_m(x) \equiv V_\lambda(x, t)$$

$$V_\lambda(x, t) = A v_\lambda(x, t; a) + B v_\lambda(x, t; b)$$

$V_\lambda(x, t)$ satisfies the difference equation (IV$_h$) which can now be written deleting the index λ (we have changed m into $m + 1$)

$$V_{h,\lambda} \begin{cases} V(x, t + \lambda) = p(x) V(x + h, t) + q(x) V(x - h, t) \\[1mm] t > 0 \quad a < x < b \\[1mm] V(x, t) \equiv A \quad \text{if} \quad x \leq a \qquad t \geq 0 \\[1mm] V(x,) \quad \equiv B \quad \text{if} \quad x \geq b \qquad t \geq 0 \\[1mm] V(x, t) \equiv 0 \quad \text{if} \quad a < x < b \quad t = 0 \end{cases}$$

This equation is of the first order in t and of the second order in x. It can hence be expressed in terms of the differences:

$$\Delta_t V(x, t) \equiv \frac{V(x, t + \lambda) - V(x, t)}{\lambda}$$

$$\Delta_x V(x, t) \equiv \frac{V(x + h, t) - V(x, t)}{h}$$

$$\Delta_{xx} V(x, t) \equiv \frac{\Delta_x V(x, t) - \Delta_x V(x - h, t)}{h}$$

We get indeed, subtracting $V(x, t)$ from both sides:

$$\lambda \Delta_t V(x, t) = p(x) V(x + h, t) + q(x) V(x - h, t) - V(x, t)$$

$$= p(x) [V(x + h, t) - V(x, t)] - q(x) [V(x, t) - V(x - h, t)]$$

$$= hp(x) \Delta_x V(x, t) - hq(x) \Delta_x V(x - h, t)$$

$$= h[p(x) - q(x)] \Delta_x V(x, t) + hq(x) [\Delta_x V(x, t) - \Delta_x V(x - h, t)]$$

$$= h[p(x) - q(x)] \Delta_x V(x, t) + h^2 q(x) \Delta_{xx} V(x, t)$$

Returning to the definition of $p(x)$ and $q(x)$, the equation takes the form:

$$\frac{\lambda}{h^2} D(x) \cdot \Delta_t V(x, t) = \beta(x) \Delta_{xx} V(x, t) + 2\alpha(x) \Delta_x V(x, t)$$

where
$$D(x) \equiv 2\beta(x) + 2h\alpha(x)$$

At a point (x, t) the solution $V(x, t)$ depends not only on the limit conditions A and B, but also on the space–time squaring we have adopted. There exist hence as many estimates of the average score $V(x, t)$ as there are possible choices of the pair (h, λ) or, this amounts to the same, of transport games $x \to x'$ restricted to a time t' less than or equal to t. These estimates present theoretically a large discrepancy since when λ decreases from t to 0, m assumes all the positive integer values and $V(x, t)$ becomes equal successively to each of the numbers $V_1(x), V_2(x), V_3(x), ..., V_m(x)$ the first of which are in general equal to zero, whereas $V(x) = \lim\limits_{m \to \infty} V_m(x)$ is finite and not equal to zero*.

Consequently:

1) when $\lambda \to 0$ for constant h, $V(x, t)$ coincides at the limit with the function $V(x)$ which *does not depend on time* and satisfies the difference equation (III$_h$):

$$\text{III}_h \quad \begin{cases} \beta(x) \Delta_{xx} V(x) + 2\alpha(x) \Delta_x V(x) = 0 & \text{if} \quad a < x < b \\ V(x) \equiv A & \text{if} \quad x \leq a \\ V(x) \equiv B & \text{if} \quad x \geq b \end{cases}$$

2) When λ and $h \to 0$, so that $\lambda/h^2 \to 0$, $V(x, t)$ converges more precisely to $U(x)$, integral of the parabolic differential equation (III)

$$\text{III} \quad \begin{cases} \beta(x) \dfrac{d^2 U}{dx^2} + 2\alpha(x) \dfrac{dU}{dx} = 0 & \text{if} \quad a < x < b \\ U(a) = A \\ U(b) = B \end{cases}$$

In both cases, the surface $S_{h, \lambda}$, $V = V(x, t)$, degenerates into a cylinder with generatrix parallel to the time axis and directrix equal to the cross-section by the plane $t = \infty$, that is $V = V(x)$, $T = $ const in the first case, and $V = U(x)$, $t = $ const in the second case.

---

* Except perhaps at the point $x \in (a, b)$, if it exists, such that
$$Av(x, a) + Bv(x, b) = 0$$

3) When $h \to 0$ for constant λ, every particle ends up by surviving a transport game restricted to a finite number, t/λ, of transitions. We have hence

$$\lim_{h \to 0} v_\lambda(x, t; a) = \lim_{h \to 0} v_\lambda(x, t; b) = \lim_{h \to 0} V(x, t) = 0$$

At the limit $(V_{h,\lambda})$ admits the trivial solution:

$$V(x, t) \equiv 0 \quad \text{if} \quad a < x < b \quad t \geq 0$$
$$V(x, t) \equiv A \quad \text{if} \quad x \leq a \qquad\quad t \geq 0$$
$$V(x, t) \equiv B \quad \text{if} \quad x \geq b \qquad\quad t \geq 0$$

4) This solution remains valid when h and $\lambda \to 0$ so that $h^2/\lambda \to 0$. In other words, no path reaches the extremities of the segment AB, no matter how large the number of transitions t/λ, if the covered space h is infinitesimal of order greater than $\sqrt{\lambda}$. This can readily be verified by referring to the difference equation which reduces, at the limit, to

$$\frac{\partial V(x, t)}{\partial t} = 0$$

whose integral is indeed $V(x, t) \equiv 0$, $a < x < b$, taking the limit conditions into account.

5) Finally, when λ and $h \to 0$ so that λ/h^2, has a finite, non-zero positive limit μ, $V(x, t)$ converges to the solution $U(x, t)$ of the parabolic partial differential equation

$$\text{VI} \quad \begin{cases} 2\mu \dfrac{\partial U}{\partial t} = \dfrac{\partial^2 U}{\partial x^2} + \dfrac{2\alpha(x)}{\beta(x)} \dfrac{\partial U}{\partial x} \\[2mm] U(a, t) \equiv A \quad t \geq 0 \\[1mm] U(b, t) \equiv B \quad t \geq 0 \\[1mm] U(x, t) \equiv 0 \quad t = 0, \quad a < x < b \end{cases}$$

This is a Fokker–Planck equation. It allows us to calculate the scattering with anisotropy for particles whose motion occurs in only one direction, inside a non-absorbing finite domain.

At each point $x \in (a, b)$, $U(x, t)$ represents the probability of vanishing of the material at one of the extremities, in a time shorter than t. More precisely, if:

$$\left. \begin{array}{l} A = 0 \\[1mm] B = 1 \end{array} \right\} \quad U(x, t) \equiv u(x, t; b)$$

denotes the percentage of leakages at B

$$\left. \begin{array}{l} A = 1 \\ \\ B = 0 \end{array} \right\} \ U(x, t) \equiv u(x, t; a)$$

if

denotes the percentages of leakages at A.

The difference $u(x, t) \equiv 1 - u(x, t; a) - u(x, t; b)$ represents the percentage of particles born at x which survive after time t. We show that

$$\lim_{t \to \infty} u(x, t) = 0$$

The integral of (VI) is a linear combination of the preceding solutions:

$$U(x, t) \equiv Au(x, t; a) + Bu(x, t; b)$$

when $t \to \infty$, $u(x, t; a) \to u(x, a)$ and $u(x, t; b) \to u(x, b)$, $U(x, t) \to U(x)$ $\equiv Au(x; a) + Bu(x; b)$, integral of the parabolic differential equation (III).

GENERALIZATION

Equation (VI) is more general than the game it describes. We can show, indeed, that there exist an infinity of transport games differing by the mechanism of the transitions but leading, at the limit, to the same flux of material through the extremities A and B. The gambler's ruin game is one of them, the simplest, because each transition only decides between *two* possible events $x \to x' = x + h$ and $x \to x' = x - h$. The development of the game corresponds to a Markov chain since the respective probabilities $p(x)$ and $q(x)$ depend only on the immediately preceding stage. The generalization consists of enlarging the spectrum (x') of possible values of the random variable x' while maintaining the fundamental character of the probability law, say, for instance, when (x') is a spectrum of beams:

$$Pr[x \to x'] = p(x, x')$$

When (x') is composed of both beams and bands, the distribution of the transitions is defined by a distribution function:

$$Pr[x \to x' \leq \xi] = F(x/\xi)$$

In both cases, the probabilities depend only on the previous stage, x.

The concept of geometrical step, h, loses some of its interest since, at each transition, the jump $|x' - x|$ can assume more than one value. On the other hand, all the transitions have same length, λ, so that the time net,

$t = m\lambda$, $m = 1, 2, 3, \cdots$ still represents the times of the successive stages of a particle born at x_0 at time $t = 0$. A second generalization consists precisely of introducing time in the probability law. Let $X(t)$ be the function which represents the abscissae of the successive stages of a path

$$X(0) \quad = x_0$$

$$X(\tau) \quad = x$$

$$X(\tau + \lambda) \ = x'$$

When a particle behaves according to a Markov chain, the probability law of a transition $X(t) = x \to X(t + \lambda) = x' \leq \xi$ depends uniquely on t and x on the one hand, on λ and ξ on the other and not, moreover, on the values X might have taken at times τ previous to t. Denoting by $F(t, x/\lambda, \xi)$ the distribution function of the distribution of the transitions we have, by definition:

$$Pr[X(t) = x \to X(t + \lambda) = x' \leq \xi] = F(t, x/\lambda, \xi)$$

which generalizes an earlier formula.

The chain is *constant* when the probability law is independent of t or, which amounts to the same, of the rank m of the stage under consideration. The distribution then depends only on three variables x, ξ and λ which we will write, emphasizing the step λ:

$$Pr[X(t) = x \to X(t + \lambda) = x' \leq \xi] = F_\lambda(x/\xi)$$

The probability of a given transition $x \to x' = \xi + 0$, up to $d\xi$, is obtained by differentiation ($d\xi > 0$)*

$$Pr[X(t) = x \to X(t + \lambda) = \xi + 0, \text{ up to } d\xi] = d_\xi F_\lambda(x/\xi)$$

In the sequel we will restrict ourselves to constant Markov chains†.

Let $v_\lambda(x, t; a)$ and $v_\lambda(x, t; b)$ be the probabilities that a particle born at x will reach respectively the media $x' \leq a$ and $x' \geq b$ in a time less than or equal to t.

---

* The random variables $X(t + \lambda)$ and $X(t + \lambda) - X(t)$ have the same conditional probability when $X(t)$ is known. We have hence also ($d\xi > 0$):

$$Pr[X(t + \lambda) - X(t) \leq \xi - x \quad \text{if} \quad X(t) = x] = F_\lambda(x/\xi)$$

and $\quad P_r[X(t + \lambda) - X(t) = \xi - x + 0 \quad \text{up to} \quad d\xi] = d_\xi F_\lambda(x/\xi) \quad \text{if} \quad X(t) = x$

† Extension to non-constant chains is not difficult.

If $x \le a$ or $x \ge b$, there is, strictly speaking, no transport. The corresponding probabilities satisfy the obvious limit conditions:

$$
\begin{cases}
x \le a \quad \text{and} \quad t \ge 0 \\
v_\lambda(x, t; a) \equiv 1 \\
v_\lambda(x, t; b) \equiv 0
\end{cases}
\quad \text{and} \quad
\begin{cases}
x \ge b \quad \text{and} \quad t \ge 0 \\
v_\lambda(x, t; a) \equiv 0 \\
v_\lambda(x, t; b) \equiv 1
\end{cases}
$$

If $x \in a]\,[b$ and $t = 0$, the particle does not have a chance of reaching
one or the other extremity of the interval. We have thus also:

$$
\begin{cases}
x \in a]\,[b \quad \text{and} \quad t = 0 \\
v_\lambda(x, t; a) \equiv 0 \\
v_\lambda(x, t; b) \equiv 0
\end{cases}
$$

If $x \in a]\,[b$ and $t > 0$, the game proceeds effectively. It consists of finding among the paths which contain no more than $m = t/\lambda$ segments those
which terminate outside the interval (a, b). After the first transition, the
abscissa of the particle is ξ and it has, at most, $t - \lambda$ seconds to reach the
target. The probability of a successful path $(x, \tau = 0) \to (\xi, \tau = \lambda) \to$
$(x' \le a, \tau = t)$ is, by the theorem of combined probabilities:

$$
v_\lambda(\xi, t - \lambda; a)\, d_\xi F_\lambda(x/\xi)
$$

Each successful stroke $x \to a$, in less than t seconds, is necessarily of the
above type, so that the total probability $v_\lambda(x, t; a)$ is obtained by integration
over ξ:

$$
\begin{cases}
v_\lambda(x, t; a) = \displaystyle\int_{-\infty}^{+\infty} v_\lambda(\xi, t - \lambda; a) \cdot d_\xi F_\lambda(x/\xi) \\
x \in a]\,[b \qquad t > 0
\end{cases}
$$

Similarly we would have:

$$
\begin{cases}
v_\lambda(x, t; b) = \displaystyle\int_{-\infty}^{+\infty} v_\lambda(\xi, t - \lambda; a) \cdot d_\xi F_\lambda(x/\xi) \\
x \in a]\,[b \qquad t > 0
\end{cases}
$$

Denote by $V_\lambda(x, t)$ the mean value of a random variable s (the score)
taking the value A, B or 0 depending on whether a particle originating from x
is absorbed at A or B, at the latest at the mth transition $(m = t/\lambda)$, or survives
this transition:

$$
V_\lambda(x, t) \equiv \langle s \rangle = A v_\lambda(x, t; a) + B v_\lambda(x, t; b)
$$

We verify, by substitution, that $V_\lambda(x, t)$ satisfies the integral system:

$$
\text{VII}
\begin{cases}
x \in a] \, [b \quad\quad t > 0 \\
\quad V_\lambda(x, t + \lambda) = \displaystyle\int_{-\infty}^{+\infty} V_\lambda(\xi, t) \, d_\xi F_\lambda(x/\xi) \\
x \in a] \, [b \quad\quad t = 0 \quad V_\lambda(x, t) \equiv 0 \\
x \leq a \quad\quad\quad t \geq 0 \quad V_\lambda(x, t) \equiv A \\
x \geq b \quad\quad\quad t \leq 0 \quad V_\lambda(x, t) \equiv B
\end{cases}
$$

which can also be written in the form

$$
\frac{V_\lambda(x, t + \lambda) - V_\lambda(x, t)}{\lambda} = \frac{1}{\lambda} \int_{-\infty}^{+\infty} [V_\lambda(\xi, t) - V_\lambda(x, t)] \cdot d_\xi F_\lambda(x/\xi)
$$

It can be shown that, under certain conditions, $V_\lambda(x, t)$ has as its limit when $\lambda \to 0$ the integral $U(x, t)$ of the parabolic partial differential system (VI).

Passing to the limit

1) We have just seen that the probability law of a transition $x \to x'$ depends not only on its origin x and on its amplitude $x' - x$ but also on its length λ. Treatment of the problem leads us to considering only those procedures in which the high-amplitude transitions become infinitely unlikely when $\lambda \to 0$, which can be briefly expressed by saying that the chain is continuous:

$$
0 = \lim_{\lambda \to 0} Pr \, \{|X(t + \lambda) - X(t)| \geq \varepsilon \quad \text{if} \quad X(t) = x\}
$$

$$
= \lim_{\lambda \to 0} \int_{|\xi - x| \geq \varepsilon} d_\xi F_\lambda(x/\xi) = 1 - \lim_{\lambda \to 0} \int_{x - \varepsilon}^{x + \varepsilon} d_\xi F_\lambda(x/\xi)
$$

2) Since the continuity condition is not strict enough, we suppose, moreover, that the high-amplitude transitions make a negligible contribution to the formation of the momenta up to third order of the random variable $\xi - x$ when $\lambda \to 0$. More precisely, the two first momenta $\langle \xi - x \rangle$ and

$\langle(\xi - x)^2\rangle$ are infinitesimally small with λ of order 1, while

$$\langle(\xi - x)^3\rangle = 0(\lambda)$$

$$M_\lambda(x) \equiv \int_{-\infty}^{+\infty} (\xi - x) \, d_\xi F_\lambda(x/\xi) = \lambda \cdot M(x) + 0(\lambda)$$

$$D_\lambda(x) \equiv \int_{-\infty}^{+\infty} (\xi - x)^2 \, d_\xi F_\lambda(x/\xi) = \lambda \cdot D(x) + 0(\lambda)$$

$$T_\lambda(x) \equiv \int_{-\infty}^{+\infty} (\xi - x)^3 \, d_\xi F_\lambda(x/\xi) = 0(\lambda)*$$

where $M(x)$ and $D(x)$ are two bounded functions of $x \in (a, b)$.

On the other hand $D(x)$ is necessarily positive and non-zero since $D_\lambda(x)$ is a second-order moment.

3) In order to be able to construct effectively from any $x \in (a, b)$ and arbitrary λ, paths which terminate at A and B, the transition probability, in one sense or the other, must not vanish at any point. This is expressed in the following manner: To each $x \in (a, b)$ and for arbitrary λ (in particular, for arbitrarily small λ) we can make correspond two pairs of positive numbers $\varepsilon_\lambda^+, \eta_\lambda^+$ and $\varepsilon_\lambda^-, \eta_\lambda^-$ independent of x, such that:

$$Pr[X(t + \lambda) - X(t) > \varepsilon_\lambda^+ \quad \text{if} \quad X(t) = x] = 1 - F_\lambda(x|x + \varepsilon_\lambda^+) \geq \eta_\lambda^+$$

$$Pr[X(t + \lambda) - X(t) \leq -\varepsilon_\lambda^- \quad \text{if} \quad X(t) = x] = F_\lambda(x|x - \varepsilon_\lambda^-) \geq \eta_\lambda^-$$

4) We finally assume that $V_\lambda(\xi, t)$ can be expanded in a Taylor series up to second order inclusively about any point $x \in (a, b)$:

$$V_\lambda(\xi, t) = V_\lambda(x, t) + (\xi - x)\frac{\partial}{\partial x} V_\lambda(x, t) + \frac{1}{2}(\xi - x)^2 \frac{\partial^2}{\partial x^2} V_\lambda(x, t)$$

$$+ (\xi - x)^3 \, \Phi_\lambda(x, t; \xi)$$

where $\Phi_\lambda(x, t; \xi)$ is a uniformly bounded function of $x \in (a, b)$, $t > 0$, $\lambda > 0$ and arbitrary ξ. In particular, $\Phi_\lambda \to 0$ when $\xi \to \pm \infty †$.

---

* By definition, the integrals are absolutely convergent.

† Because of the limit conditions:

if $\qquad \xi \leq a \quad \Phi_\lambda(x, t; \xi) \equiv [A - V_\lambda(x, t)](\xi - x)^{-3}$

$$- \frac{\partial V_\lambda}{\partial x}(\xi - x)^{-2} - \frac{1}{2}\frac{\partial^2 V_\lambda}{\partial x^2}(\xi - x)^{-1}$$

if $\qquad \xi \leq b \quad \Phi_\lambda(x, t; \xi) \equiv [B - V_\lambda(x, t)](\xi - x)^{-3}$

$$- \frac{\partial V_\lambda}{\partial x}(\xi - x)^{-2} - \frac{1}{2}\frac{\partial^2 V_\lambda}{\partial x^2}(\xi - x)^{-1}$$

Substituting in (VII) and passing to the limit for $\lambda \to 0$ we get, taking (2) into account:

$$\lim_{\lambda=0} \frac{\partial V_\lambda(x, t)}{\partial t} = \lim_{\lambda=0} \left\{ M(x) \frac{\partial V_\lambda(x, t)}{\partial x} + \frac{1}{2} D(x) \frac{\partial^2 V_\lambda(x, t)}{\partial x^2} \right\}$$

which shows that any solution of (VII) satisfies, at the limit for $\lambda \to 0$, and uniformly for all $x \in (a, b)$ the parabolic partial differential equation:

$$\frac{\partial U(x, t)}{\partial t} = M(x) \frac{\partial U(x, t)}{\partial x} + \frac{1}{2} D(x) \frac{\partial^2 U(x, t)}{\partial x^2}$$

In particular, for

$$\left\{ \begin{array}{l} D(x) = \dfrac{1}{\mu} \\[2ex] M(x) = \dfrac{1}{\mu} \dfrac{\alpha(x)}{\beta(x)} \end{array} \right.$$

where μ is an arbitrary positive constant, we obtain the system (VI). Hence there do indeed exist an infinity of transport games with time constant λ satisfying at the limit, for $\lambda \to 0$, a given parabolic partial differential equation. These games are defined by the distribution functions $F_\lambda(x/\xi)$ and the conditions 2), 3).

Levitan and Petrovski gave a rigorous and rather lengthy proof of this theorem (1958). We merely give the preceding indications and the following statement:

A particle with abscissa $X(t)$ follows a rectilinear path on the x-axis by successive jumps of constant time λ and random amplitude $X(t + \lambda) - X(t)$ defined by the conditional distribution functions

$$F_\lambda(x/\xi) \equiv Pr[X(t + \lambda) - X(t) \leq \xi - x \quad \text{if} \quad X(t) = x]$$

We denote by $V_\lambda(x, t)$ the mean value of a random variable taking the value A, B or 0 depending on whether a particle originating from $x \in (a, b)$ at time $t = 0$ is absorbed at $x' \leq a$ or $x' \geq b$ at the latest at the mth transition ($m = t/\lambda$), or survives this transition.

Theorem 1   a) If the moments of the random variable $X(t + \lambda) - X(t)$ *exist up to the third order for arbitrary* $X(t) = x \in (a, b)$ *and are of the form:*

$$\begin{cases} M_\lambda(x) \equiv \int\limits_{-\infty}^{+\infty} (\xi - x)\, d_\xi F_\lambda(x/\xi) = \lambda k(x)\, \alpha(x) + 0(\lambda) \\[2mm] D_\lambda(x) \equiv \int\limits_{-\infty}^{+\infty} (\xi - x)^2\, d_\xi F_\lambda(x/\xi) = \lambda k(x)\, \beta(x) + 0(\lambda) \\[2mm] T_\lambda(x) \equiv \int\limits_{-\infty}^{+\infty} (\xi - x)^3\, d_\xi F_\lambda(x/\xi) = 0(\lambda) \end{cases}$$

uniformly for all $x \in (a, b)$

where $\alpha(x)$, $\beta(x)$ *and* $k(x)$ *are continuous functions of* x *and* $\beta(x)$ *and* $k(x)$ *are positive.*

  b) If there exist two pairs of positive numbers ε_λ^+, η_λ^+ *and* ε_λ^-, η_λ^- *independent of* x *and such that*

$$1 - F_\lambda(x/x + \varepsilon_\lambda^+) \geqq \eta_\lambda^+ \quad F_\lambda(x/x - \varepsilon_\lambda^-) \geqq \eta_\lambda^-$$

for any $\lambda \leqq \lambda_0$, *where* λ_0 *is a positive number independent of* x.

  The limit of $V_\lambda(x, t)$ *when* $\lambda \to 0$ *is the integral* $U(x, t)$ *of the parabolic partial differential equation*

$$\frac{\partial U}{\partial t} = \frac{1}{2} k(x)\, \beta(x)\, \frac{\partial^2 U}{\partial x^2} + k(x)\, \alpha(x) \cdot \frac{\partial U}{\partial x}$$

satisfying the limit conditions:

$$U(a, t) \equiv A \quad t \geqq 0$$
$$U(b, t) \equiv B \quad t \geqq 0$$
$$U(x, 0) \equiv 0 \quad x \in a]\,[b$$

Application   The previous theorem gives sufficient conditions for convergence in a Markov process when the relaxation time tends to zero.

  We are going to apply it to the gambler's ruin problem. We will find again the condition $\lambda = \mu h^2$ connecting the relaxation time and the geometrical step.

  The distribution function and its "density" have expression

$$F_\lambda(x/\xi) \equiv q(x)\, Y(\xi - x + h) + p(x)\, Y(\xi - x - h)$$
$$d_\xi F_\lambda(x/\xi) \equiv \{q(x)\, \delta(\xi - x + h) + p(x)\, \delta(\xi - x - h)\}\, d\xi$$

where $Y(u)$ and $\delta(u)$ denote respectively the echelon function and the Dirac measure. $F_\lambda(x/\xi)$ is represented in the diagram below:

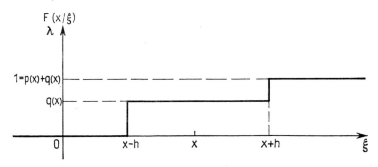

Figure 30

$$F_\lambda(x/\xi) = Pr[X(t + \lambda) \leq \xi \quad \text{if} \quad X(t) = x]$$

$$p(x) = \frac{1}{D(x)} \{\beta(x) + 2h\alpha(x)\}$$

$$q(x) = \frac{1}{D(x)} \beta(x)$$

$$D(x) = 2\beta(x) + 2h\alpha(x)$$

Having established this, we now calculate the three first moments of the random variable $X(t + \lambda) - X(t)$:

$$M_\lambda(x) = hp(x) - hq(x) = h^2 \frac{\alpha(x)}{\beta(x)} + 0(h^2)$$

$$D_\lambda(x) = h^2 p(x) + h^2 q(x) = h^2$$

$$T_\lambda(x) = h^3 p(x) - h^3 q(x) = h^4 \frac{\alpha(x)}{\beta(x)} + 0(h^4)$$

To verify condition (a) of the theorem, it suffices to adjust the amplitude of the jumps to the length of the relaxations by setting:

$$\left\{ \begin{array}{l} \lambda = \mu h^2 \\[2ex] k(x) = \dfrac{1}{\mu \beta(x)} \end{array} \right.$$

where μ is an arbitrary positive number. And we find thus the given condition.

Condition b) is also satisfied since the transition probabilities $p(x)$ and $q(x)$ do not vanish at any point of the interval as soon as h is sufficiently small. More precisely:

$$1 - F_\lambda\left(x\left|x + \frac{h}{2}\right.\right) = p(x) = \frac{1}{2}\left(1 + h\frac{\alpha(x)}{\beta(x)}\right) + 0(h)$$

$$F_\lambda\left(x\left|x - \frac{h}{2}\right.\right) = q(x) = \frac{1}{2}\left(1 - h\frac{\alpha(x)}{\beta(x)}\right) + 0(h)$$

Since $\dfrac{\alpha(x)}{\beta(x)}$ is bounded for any $x \in (a, b)$, we can always choose $\lambda \equiv \mu h^2$ sufficiently small so that $p(x)$ and $q(x)$ remain larger than a positive quantity in the interval (a, b). We have:

$$\begin{cases} \varepsilon_\lambda^+ = \dfrac{h}{2} = \dfrac{1}{2}\sqrt{\dfrac{\lambda}{\mu}} \\[2mm] \eta_\lambda^+ = 1 - \dfrac{1}{2}\,\dfrac{1}{\min\left\{1 + \sqrt{\dfrac{\lambda}{\mu}}\dfrac{\alpha(x)}{\beta(x)}\right\}} \end{cases}$$

and

$$\begin{cases} \varepsilon_\lambda^- = \dfrac{h}{2} = \dfrac{1}{2}\sqrt{\dfrac{\lambda}{\mu}} \\[2mm] \eta_\lambda^- = \dfrac{1}{2}\,\dfrac{1}{\max\left\{1 + \dfrac{\sqrt{\lambda}}{\sqrt{\mu}}\dfrac{\alpha(x)}{\beta(x)}\right\}} \end{cases}$$

Consequently, when h and $\lambda = \mu h^2$ tend to zero, the limit of $V_\lambda(x, t)$ is indeed the integral $U(x, t)$ of the partial differential equation

$$2\mu\frac{\partial U}{\partial t} = \frac{\partial^2 U}{\partial x^2} + 2\frac{\alpha(x)}{\beta(x)}\frac{\partial U}{\partial x}$$

satisfying the limit conditions:

$$\begin{cases} U(a, t) \equiv A & t \geqq 0 \\ U(b, t) \equiv B & t \geqq 0 \\ U(x, 0) \equiv 0 & x \in a]\,[b \end{cases}$$

GENERAL DIRICHLET PROBLEMS FOR DIFFERENTIAL AND DIFFERENCE EQUATIONS

Generalization to *several dimensions* of the preceding results will be considered only for two dimensions.

Problem (D)

Consider a Jordan curve C, with interior R. (We denote by $\bar{R}$ the union $R \bigcup C$.) We take a continuous function $\Phi(x, y)$ on C. To find $U(x, y)$, the solution of the elliptic partial differential equation:

$$L(U) \equiv \beta_{11} \frac{\partial^2 U}{\partial x^2} + 2\beta_{12} \frac{\partial^2 U}{\partial x \partial y} + \beta_{22} \frac{\partial^2 U}{\partial y^2} + 2\alpha_1 \frac{\partial U}{\partial x} + 2\alpha_2 \frac{\partial U}{\partial y} = 0 \quad (1)$$

and taking the values $\Phi(x, y)$ on C.

We suppose:

a) the α and β are continuous in x and y, have first and second derivatives, in a region containing R and in $\bar{R}$.

b) $\beta_{11} > 0, \beta_{22} > 0, \beta_{11}\beta_{22} - \beta_{12}^2 > 0$.

We then consider a net

$$\begin{cases} x = x_0 + jh \\ y = y_0 + jh \end{cases} \quad j = 0, \pm 1, \pm 2 \text{ etc.}$$

where x_0, y_0 are suitably chosen, and h is a constant (the step), which defines a field R_h.

We note that:

a) Some neighbourhoods of points of R_h are not in R_h

b) For any pair (x, y) of R_h, the pair $(x + h, y + h)$ is not necessarily contained in R_h.

We denote by C_h the set of these neighbourhoods and of these points exterior to R_h; and by $\bar{R}_h$ the union $R_h \bigcup C_h$.

Consider the first and second differences:

$$\Delta_x V = \frac{V(x + h, y) - V(x, y)}{h}$$

$$\Delta_{xx} V = \frac{V(x + h, y) + V(x - h, y) - 2V(x, y)}{h^2}$$

$$\Delta_{xy} V = \frac{V(x + h, y + h) - V(x + h, y) - V(x, y + h) + V(x, y)}{h^2}$$

and the analogous formulas for $\Delta_y V$ and $\Delta_{yy} V$.

For simplicity, we denote the pair (x, y) by a letter P or Q and we attach the indices (1) to (5) to the neighbouring points of P as indicated in the diagram.

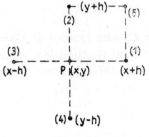

Figure 31

We then consider *the difference operator:*

$$L(V) = \beta_{11}\Delta_{xx}V + 2\beta_{12}\Delta_{xy}V + \beta_{22}\Delta_{yy}V + 2\alpha_1\Delta_xV + 2\alpha_2\Delta_yV \quad (2)$$

where the coefficients have the same properties as for (1).

Problem (D_h)

We take $\Phi(x, y)$ on C_h. Solve

$$L(V) = 0 \quad (3)$$

where $V \equiv \Phi$ on C_h.

We set

$$p_1(P) = \frac{\beta_{11} - 2\beta_{12} + 2h\alpha_1}{D} \quad p_2(P) = \frac{\beta_{22} - 2\beta_{12} + 2h\alpha_2}{D}$$

$$p_3(P) = \frac{\beta_{11}}{D} \qquad p_4(P) = \frac{\beta_{22}}{D} \qquad p_5(P) = \frac{2\beta_{12}}{D}$$

where:

$$D = D(P) = 2\beta_{11} + 2\beta_{22} - 2\beta_{12} + 2h(\alpha_1 + \alpha_2)$$

then equation (3) is written:

$$\boxed{V(P) = \sum_1^5 p_k(P) V(P_k)} \quad (4)$$

We also have $\sum_1^5 p_j(P) = 1$, but here p_1 or p_2 can be negative if:

$$2\beta_{12} > \beta_{11} \quad \text{or} \quad 2\beta_{12} > \beta_{22}$$

In the sequel we suppose:

$$2\beta_{12} < \beta_{11} \quad \text{and} \quad 2\beta_{12} < \beta_{22}$$

for P in R_h, and *also* that h is sufficiently small so that, if α_1 and α_2 are negative, the $p_j(P)$ remain non-negative in R_h.

Problem   Let Q_j $(j = 1, 2, \cdots)$ be the points of C_k. A particle leaves a point P_0 of R_h and describes a random path on R_h with *the conditions* that at P the probabilities of going to P_k $(k = 1, 2, \cdots, 5)$ are $p_k(P)$. Then the particle reaches a point Q_j, we score $\Phi(Q_j)$.
What is the mean value of the score?
   We set:

$$\Phi_0 = |\text{Inf } \Phi(Q_j)|_{Q_j \in C_h}$$

which implies that:

$$\Phi(Q_j) + \Phi_0 \geq 0$$

The mean score, *if it exists,* is the difference of the corresponding mean values of the scores $\Phi(Q_j) + \Phi_0$ at Q_j and Φ_0 at Q_j *provided* that these mean values *exist. It suffices hence* to study the case of a non-negative set of values $\psi(Q_j)$ on the boundary.

Fundamental result   Consider a path with at most m stages; $V_m(P)$ is the value of the $\psi(Q_j)$ if we actually reach a point Q_j, zero if we are still in R_h. For any P in R_h, we have:

$$\left. \begin{aligned} V_m(P) &= \sum_1^5 p_k(P)\, V_{m-1}(P_k) \\ V_m(Q_j) &= \psi(Q_j) \quad j = 1, 2 \ldots \end{aligned} \right\} \tag{5}$$

Hence:

$$V_{m+1}(P) - V_m(P) = \sum_1^5 p_k(P)\,[V_m(P_k) - V_{m-1}(P_k)]$$

And hence if, for a value of m, we have $V_{m+1} \geq V_m$ for any P in R_h, this will also be true for higher values of m. Now $V_0(P) = 0$ for any P in R_h and $V_0(Q_j) = \psi(Q_j)$. Since the p_k are non-negative, we have hence:

$$V_1(P) \geq V_0(P) \quad \text{for any } P \text{ in } R_h \text{ and}$$

$$0 \leq V_m \leq \sup_j \psi(Q_j).$$

$\{Y_m\}$ is hence a bounded monotonic sequence, and admits a limit for any P in R_h. Passing to the limit in (5), we see that the limit $V(P)$ satisfies the equations (3) and (4). We can show asymptotically (see the preceding problems) that if $m \to \infty$, the probability of zero score is zero.

So that finally: *The mean of the scores of the paths for an unlimited amount of stages gives a solution of the problem* (D_h).

Remark   Under certain conditions (cf. Theorem 1 in the one-dimensional case) $V(P)$ tends, when h tends to zero, to the solution of the problem (D). The proofs and the results generalize Theorem 1 (we have convergence to the solution when $\lambda \to 0$), since the preceding random paths satisfy the hypotheses of the theorem.